Geometry

Applying • Reasoning • Measuring

Chapter 2 Resource Book

The Resource Book contains the wide variety of blackline masters available for Chapter 2. The blacklines are organized by lesson. Included are support materials for the teacher as well as practice, activities, applications, and assessment resources.

McDougal Littell
A HOUGHTON MIFFLIN COMPANY
Evanston, Illinois • Boston • Dallas

Contributing Authors

The authors wish to thank the following individuals for their contributions to the Chapter 2 Resource Book.

Eric J. Amendola
Patrick M. Kelly
Edward H. Kuhar
Lynn Lafferty
Dr. Frank Marzano
Wayne Nirode
Dr. Charles Redmond
Paul Ruland

ISBN: 0-618-02065-9

23456789-VEI- 04 03 02 01 00

Contents

2 *Reasoning and Proof*

Contents

Contents

Descriptions of Resources

This Chapter Resource Book is organized by lessons within the chapter in order to make your planning easier. The following materials are provided:

Tips for New Teachers These teaching notes provide both new and experienced teachers with useful teaching tips for each lesson, including tips about common errors and inclusion.

Parent Guide for Student Success This guide helps parents contribute to student success by providing an overview of the chapter along with questions and activities for parents and students to work on together.

Prerequisite Skills Review Worked-out examples are provided to review the prerequisite skills highlighted on the Study Guide page at the beginning of the chapter. Additional practice is included with each worked-out example.

Strategies for Reading Mathematics The first page teaches reading strategies to be applied to the current chapter and to later chapters. The second page is a visual glossary of key vocabulary.

Lesson Plans and Lesson Plans for Block Scheduling This planning template helps teachers select the materials they will use to teach each lesson from among the variety of materials available for the lesson. The block-scheduling version provides additional information about pacing.

Warm-Up Exercises and Daily Homework Quiz The warm-ups cover prerequisite skills that help prepare students for a given lesson. The quiz assesses students on the content of the previous lesson. (Transparencies also available)

Activity Support Masters These blackline masters make it easier for students to record their work on selected activities in the Student Edition.

Alternative Lesson Openers An engaging alternative for starting each lesson is provided from among these four types: ***Application, Activity, Geometry Software,*** or ***Visual Approach.*** (Color transparencies also available)

Technology Activities with Keystrokes Keystrokes for Geometry software and calculators are provided for each Technology Activity in the Student Edition, along with alternative Technology Activities to begin selected lessons.

Practice A, B, and C These exercises offer additional practice for the material in each lesson, including application problems. There are three levels of practice for each lesson: A (basic), B (average), and C (advanced).

Contents

Reteaching with Additional Practice These two pages provide additional instruction, worked-out examples, and practice exercises covering the key concepts and vocabulary in each lesson.

Quick Catch-Up for Absent Students This handy form makes it easy for teachers to let students who have been absent know what to do for homework and which activities or examples were covered in class.

Cooperative Learning Activities These enrichment activities apply the math taught in the lesson in an interesting way that lends itself to group work.

Interdisciplinary Applications/Real-Life Applications Students apply the mathematics covered in each lesson to solve an interesting interdisciplinary or real-life problem.

Math and History Applications This worksheet expands upon the Math and History feature in the Student Edition.

Challenge: Skills and Applications Teachers can use these exercises to enrich or extend each lesson.

Quizzes The quizzes can be used to assess student progress on two or three lessons.

Chapter Review Games and Activities This worksheet offers fun practice at the end of the chapter and provides an alternative way to review the chapter content in preparation for the Chapter Test.

Chapter Tests A, B, and C These are tests that cover the most important skills taught in the chapter. There are three levels of test: A (basic), B (average), and C (advanced).

SAT/ACT Chapter Test This test also covers the most important skills taught in the chapter, but questions are in multiple-choice and quantitative-comparison format. (See *Alternative Assessment* for multi-step problems.)

Alternative Assessment with Rubrics and Math Journal A journal exercise has students write about the mathematics in the chapter. A multi-step problem has students apply a variety of skills from the chapter and explain their reasoning. Solutions and a 4-point rubric are included.

Project with Rubric The project allows students to delve more deeply into a problem that applies the mathematics of the chapter. Teacher's notes and a 4-point rubric are included.

Cumulative Review These practice pages help students maintain skills from the current chapter and preceding chapters.

CHAPTER 2

Tips for New Teachers

For use with Chapter 2

Chapter Support

LESSON 2.1

TEACHING TIP To help students visualize Postulates 5 through 11 in Example 5 on page 73, draw a separate diagram for each one and have students do the same in their notes. Show students how to draw three-dimensional planes using the labeling conventions used in the textbook. Students should gain more confidence in their ability to draw accurate figures and to analyze them.

INCLUSION To help students with limited English proficiency, use classroom examples to support the solution diagrams in Example 7 on page 74. Students could consider the plane of the ceiling and the plane of a wall for part (a). For part (b), points A, B, and C might be on the front board and point D might be located at a student's desk. Two rulers or two pencils could be used to represent lines m and n in part (c). Skew lines are defined in Lesson 3.1; you might want to mention them here. Ask students to look at the intersection of the ceiling and a wall. Then ask them to find another line in the classroom that is neither parallel to nor coplanar with the first line they found.

LESSON 2.2

TEACHING TIP It is important to emphasize that a conditional statement and its converse are both true for all the definitions used in the study of geometry. However, students need to keep in mind the Study Tip on page 81, which indicates that this idea does not apply to all postulates.

TEACHING TIP To help students understand and feel comfortable with Example 2 on page 80, have them write the conditional statement and its converse for the definition of vertical angles and then write the biconditional statement. The same may be done for the definition of linear pair.

LESSON 2.3

TEACHING TIP An understanding of the differences between inductive and deductive reasoning is not always clear to students. Following Example 3 on page 89, you might have students identify situations where they reached a conclusion or made a conjecture based on each type of reasoning. This should emphasize that inductive reasoning is based on past observations or patterns and that deductive reasoning is based on accepted statements or facts.

COMMON ERROR In Exercises 36 through 42 on page 93, students may have trouble understanding the notation used, since the subscript of p is not the same as the numbered angle. Point out that the symbolic logic notation is given first, followed by a colon, and then the written statement about the angles is shown.

LESSON 2.4

INCLUSION To refresh students' memory of the algebraic properties of equality and to help students with limited English proficiency, have students give several numeric examples of each property. This can be done prior to Example 1 on page 96.

LESSON 2.5

TEACHING TIP Theorems are introduced in this lesson and a numbering system is applied to them. Students should be reminded that like postulates, theorems are numbered to make their textbook reference easier. Stress that students must know the statement of the theorem and that using only the theorem number is not acceptable. Some theorems are named and you must decide when to have students write out the statement of the theorem rather than using just the name. Consider letting students use just the name once the chapter is completed and they fully understand the theorem.

TEACHING TIP Two column proofs are introduced here. Prior to the proof in Example 1 on page 102, have students review the definition of congruent segments as a biconditional statement. Students must be aware of what can be used as a statement and as a reason when writing two-column proofs.

TEACHING TIP Once the transitive property is understood, it may be used in a slightly different format than was presented earlier (in Lesson 2.4 on page 96). This is evident in Example 2 on page 103, where $LK = 5$ and $JK = 5$ leads to a statement that $LK = JK$ by the transitive property of equality. This format occurs in some other examples in the textbook as well. Students need to

CHAPTER 2 CONTINUED

Tips for New Teachers

For use with Chapter 2

be reassured that "if $a = b$ and $b = c$, then $a = c$" is equivalent to "if $a = b$ and $c = b$, then $a = c$." You could point out that changing $b = c$ to $c = b$ is justified by the symmetric property of equality.

LESSON 2.6

TEACHING TIP Prior to the proof in Example 1 on page 109, have students review the definition of congruent angles as a biconditional statement. Be aware that the standard format of the transitive property of equality is used in the proofs in Example 1 on page 109, Example 2 on page 110, and Example 5 on page 112. However, the changed format for the transitive property is used in the proofs in Example 3 on page 110 and Example 4 on page 111. Again, if a student is confused by Examples 3 and 4 you could have them add the step justified by the symmetric property. However, be careful about the level of rigor you demand. Students sometimes resist proving what is obvious to them.

Outside Resources

BOOKS/PERIODICALS

Rothbart, Andrea. "Learning to Reason from Lewis Carroll." *Mathematics Teacher* (January 1998); pp. 6–10, 96.

McGivney, Jean M. and Thomas C. DeFranco. "Geometry Proof Writing: A Problem-Solving Approach a la Polya." *Mathematics Teacher* (October 1995); pp. 552–555.

ACTIVITIES/MANIPULATIVES

Picciotto, Henry. *Geometry Labs*. Collection of 78 blackline masters to use with manipulatives to promote reasoning skills. Emeryville, CA. Key Curriculum Press.

SOFTWARE

Rethinking Proof with The Geometer's Sketchpad. Verify, explain, explore, and communicate proof. Berkeley, CA; Key Curriculum Press.

NAME ______________________________ DATE ____________

Parent Guide for Student Success

For use with Chapter 2

Chapter Overview One way that you can help your student succeed in Chapter 2 is by discussing the lesson goals in the chart below. When a lesson is completed, ask your student to interpret the lesson goals for you and to explain how the mathematics of the lesson relates to one of the key applications listed in the chart.

Lesson Title	*Lesson Goals*	*Key Applications*
2.1: Conditional Statements	Recognize and analyze a conditional statement. Write postulates about points, lines, and planes using conditional statements.	• Flowers • Advertising • Research Buggy
2.2: Definitions and Biconditional Statements	Recognize and use definitions and biconditional statements.	• Snow Leopards • Musical Groups • Winds at Sea
2.3: Deductive Reasoning	Use symbolic notation to represent logical statements. Form conclusions by applying the laws of logic to true statements.	• Civil Rights Memorial • Kolob Arch • Robotics
2.4: Reasoning with Properties from Algebra	Use properties from algebra. Use properties of length and measure to justify segment and angle relationships.	• Fitness • Auto Racing • Pay Raises
2.5: Proving Statements about Segments	Justify statements about congruent segments. Write reasons for steps in a proof.	• Bridges • Carpentry • Optical Illusion
2.6: Proving Statements about Angles	Use angle congruence properties. Prove properties about special pairs of angles.	• Scissors • Wall Trim • Picture Frames

Test-Taking Strategy

Making sure you understand directions is the test-taking strategy featured in Chapter 2 (see page 122). Be sure your student is familiar with directions that are likely to be on a test. It is important to answer the right question and to write answers in the form requested. Encourage your student to listen carefully to spoken directions at the beginning of a test and to ask questions about any directions, written or spoken, that are not understood.

CHAPTER 2 CONTINUED

Chapter Support

NAME ______________________________ DATE ____________

Parent Guide for Student Success

For use with Chapter 2

Key Ideas Your student can demonstrate understanding of key concepts by working through the following exercises with you.

Lesson	Exercise
2.1	Rewrite the conditional statement in if-then form. Then decide whether the statement is *true* or *false*. If false, provide a counterexample. "It is winter in December."
2.2	Rewrite the true statement in if-then form and write the converse. If the converse is true, combine it with the if-then statement to form a true biconditional statement. If the converse is false, provide a counterexample. "Two perpendicular lines form right angles."
2.3	Use the Law of Syllogism to write the statement that follows from the pair of true statements. If Roger earns money, he buys new clothes. If Roger works, he earns money.
2.4	Use the transitive property of equality to complete the statement. If $m\angle K = 90^\circ - m\angle J = m\angle M$, then ____________.
2.5	Let H be a point in your home, W be a point where you work, and S be a point in your student's school. Suppose $HW = HS$. How could you prove $\overline{HW} \cong \overline{HS}$?
2.6	Make a sketch using the given information. Then, state all of the pairs of congruent angles. $\angle ABC$ and $\angle CBD$ are adjacent and supplementary. $\angle CBD$ and $\angle DBE$ are also adjacent and supplementary.

Home Involvement Activity

Directions: Think of three or four folk beliefs. For example, some people believe that it is going to rain when grandpa's arthritis hurts. State each belief as a conditional statement. State each converse. If the statement is true, must the converse also be true? If so, provide a biconditional statement. If possible, find a counterexample for each belief. If not, discuss how you could prove the statement is true.

Answers

2.1: If it is December, then it is winter; false; *Sample answer:* It is summer in December in Australia. **2.2:** If two lines are perpendicular, then they form right angles; If two lines form right angles, then they are perpendicular; true; Two lines are perpendicular if and only if they form right angles. **2.3:** If Roger works, he buys new clothes. **2.4:** $m\angle K = m\angle M$ **2.5:** by the definition of congruent segments **2.6:** $\angle ABC \cong \angle DBE$ and $\angle CBD \cong \angle ABE$

NAME ______________________ DATE ____________

Prerequisite Skills Review

For use before Chapter 2

EXAMPLE 1 Identifying Collinear and Coplanar Points

Use the diagram at the right to name the point(s).

a. A second point collinear with point A

b. A third point coplanar with points E and F

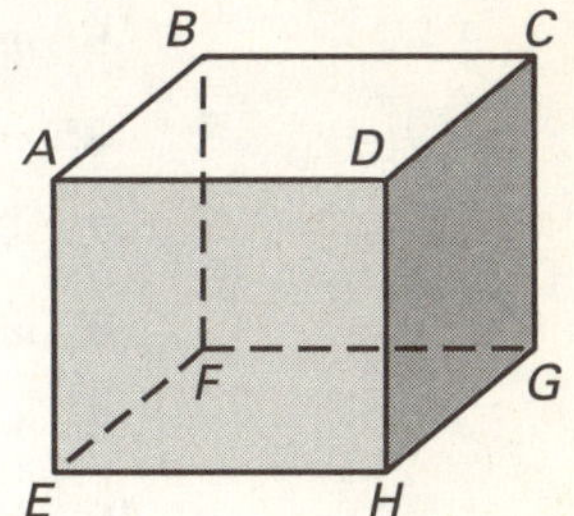

SOLUTION

a. Points A and B lie on the same line, so they are collinear. Also, D is collinear with A, and E is collinear with A.

b. There are many correct answers. For instance, points A, E, and F lie on the same plane, so they are coplanar.

Exercises for Example 1

Use the diagram at the right to name the point(s).

1. A second point collinear with point T

2. A third point coplanar with points Y and Z

3. A third point coplanar with points T and X

4. A fourth point coplanar with points W, S, and V

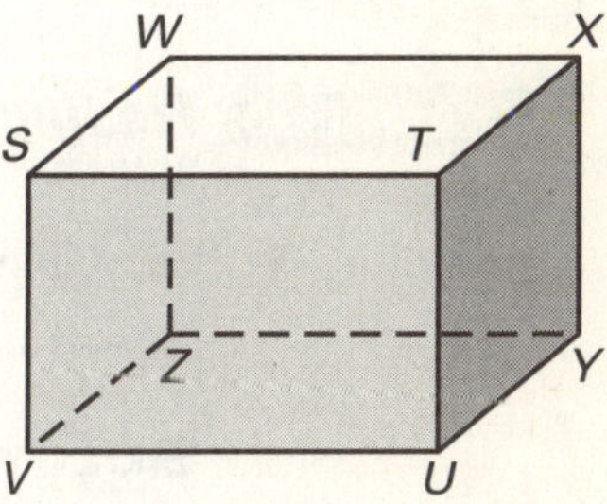

EXAMPLE 2 Finding Angle Measures

In the diagram on the right, ∠2 has a measure of 40°. Find the measure of the angle.

a. $m\angle 1$ **b.** $m\angle 4$

SOLUTION

a. ∠1 and ∠2 are a linear pair. So, the sum of their measures is 180°.

$$m\angle 1 + m\angle 2 = 180^\circ$$
$$m\angle 1 + 40 = 180^\circ$$
$$m\angle 1 = 140^\circ$$

b. ∠2 and ∠4 are vertical angles. So, they are congruent and have the same measure.

$$m\angle 4 = m\angle 2 = 40^\circ$$

NAME ______________________________ DATE __________

Prerequisite Skills Review

For use before Chapter 2

Exercises for Example 2

In the diagrams below, $\angle 2$ has a measure of 54° and $\angle 5$ has a measure of 35°. Find the measure of the angle.

5. $m\angle 1$

6. $m\angle 3$

7. $m\angle 4$

8. $m\angle 6$

9. $m\angle 7$

10. $m\angle 8$

11. $m\angle 9$

12. $m\angle 10$

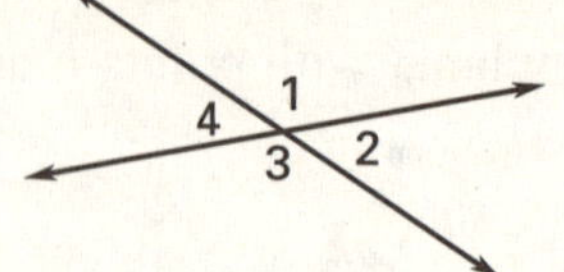

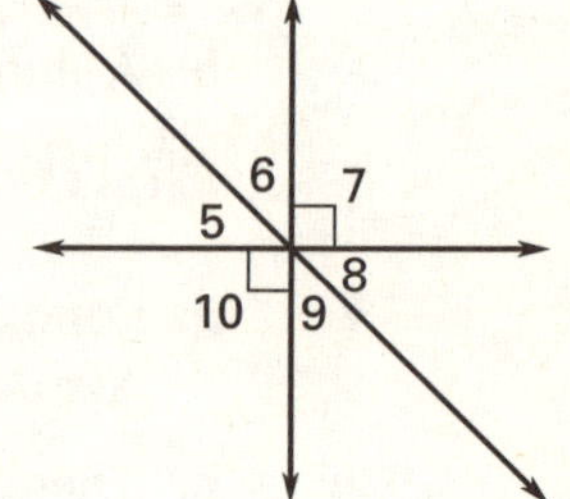

EXAMPLE 3 Finding Measures of Complements and Supplements

a. Given that $\angle B$ is a complement of $\angle A$ and $m\angle B = 45°$, find $m\angle A$.

b. Given that $\angle C$ is a supplement of $\angle D$ and $m\angle D = 70°$, find $m\angle C$.

Solution

a. $m\angle A = 90° - m\angle B = 90° - 45° = 45°$

b. $m\angle C = 180° - m\angle D = 180° - 70° = 110°$

Exercise for Example 3

13. In the table, assume that $\angle 1$ and $\angle 2$ are complementary and $\angle 1$ and $\angle 3$ are supplementary. Copy and complete the table.

$m\angle 1$	36°	54°	12°	71°	68°	113°
$m\angle 2$						
$m\angle 3$						

NAME ______________________ DATE __________

Strategies for Reading Mathematics

For use with Chapter 2

Strategy: Reading Vocabulary and Taking Notes in Geometry

In Algebra you learned many algebraic terms. Now that you are taking geometry, you will need to understand many geometric terms. Important vocabulary terms you need to learn are printed in **heavy type like this.** Take notes about these terms and examples in your notebook. The notebook below shows a sample of the notes you might take about the vocabulary terms in the next paragraph.

Two angles are **adjacent** if they share a common vertex and side but have no common interior points. Two adjacent angles are a **linear pair** if their noncommon sides are opposite rays. Two angles are **vertical angles** if their sides form two pairs of opposite rays.

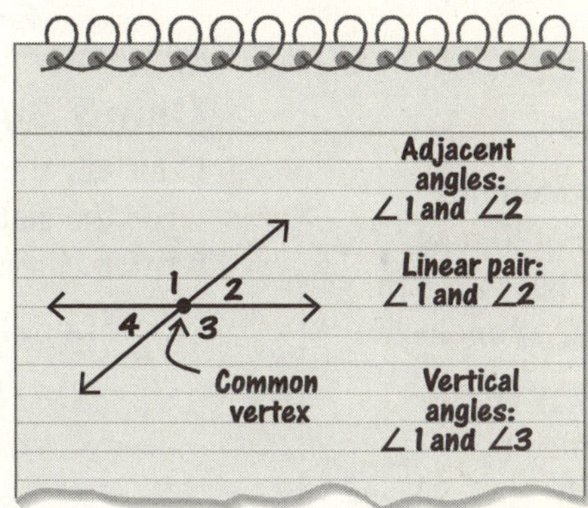

Study Tip
Reading Vocabulary

Look for terms printed in heavy type. Read the sentence that contains the term to study its definition. Then read the entire paragraph containing the sentence to be sure you understand the entire definition.

Study Tip
Take Notes

Take brief but complete notes. Write key words from the definition of each term to help you remember the meaning of the term. Draw labeled pictures or diagrams to illustrate the term. You can also include brief examples when appropriate.

Questions

1. Name other angles shown in the notebook that are adjacent. Explain how you know they are adjacent.
2. Name other angles shown in the notebook that are linear pairs. Explain how you know they are linear pairs.
3. Name other angles shown in the notebook that are vertical angles. Explain how you know they are vertical angles.
4. Which angles in the figure at the right are vertical angles? How can you use the drawing in the notebook to help you decide?

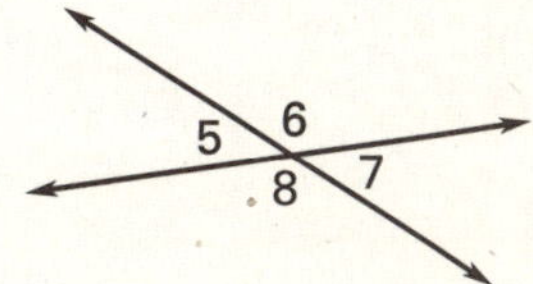

NAME ______________________________ DATE ____________

Strategies for Reading Mathematics

For use with Chapter 2

Visual Glossary

The Study Guide on page 70 lists the key vocabulary for Chapter 2 as well as review vocabulary from the previous chapter. Use the page references on page 70 or the Glossary in the textbook to review key terms from the prior chapter. Use the visual glossary below to help you understand some of the key vocabulary in Chapter 2. You may want to copy these diagrams into your notebook and refer to them as you complete the chapter.

GLOSSARY

conditional statement (p. 71) A type of logical statement that has two parts, a hypothesis and a conclusion.

converse (p. 72) The statement formed by switching the hypothesis and conclusion of a conditional statement.

inverse (p. 72) The statement formed when you negate the hypothesis and conclusion of a conditional statement.

contrapositive (p. 72) The statement formed when you negate the hypothesis and conclusion of the converse of a conditional statement.

theorem (p. 102) A true statement that follows as a result of other true statements.

two-column proof (p. 102) A type of proof written as numbered statements and reasons that show the logical order of an argument.

Using Symbolic Notation

Consider the conditional statement "If *p,* then *q*." Another way to write the statement is "*p* implies *q*." This statement can also be written using symbolic notation.

hypothesis		**conclusion**
If *p*,	then	*q*.
p	implies	*q*.
p	$\rightarrow$	*q*

The converse, inverse, and contrapositive can also be written using symbolic notation.

Converse: $q \rightarrow p$

Inverse: $\sim p \rightarrow \sim q$ (~ means "not")

Contrapositive: $\sim q \rightarrow \sim p$

Proofs

Two-column proofs can be used to prove theorems and other statements.

Given $m\angle A = 30°$
$m\angle B = 30°$

Prove $\angle A \cong \angle B$

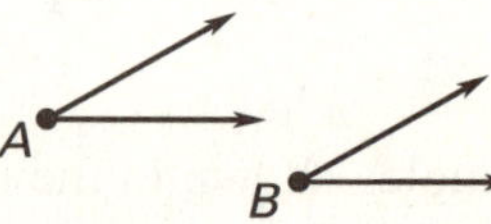

Statements	**Reasons**
1. $m\angle A = 30°, m\angle B = 30°$	1. Given
2. $m\angle A = m\angle B$	2. Substitution Property of Equality
3. $\angle A \cong \angle B$	3. Definition of congruent angles

LESSON 2.1

TEACHER'S NAME ______________ CLASS ______ ROOM ______ DATE ______

Lesson Plan

1-day lesson (See *Pacing the Chapter,* TE pages 68C–68D) For use with pages 71–78

1. Recognize and analyze a conditional statement.
2. Write postulates about points, lines, and planes using conditional statements.

State/Local Objectives ____

____

✓ Check the items you wish to use for this lesson.

STARTING OPTIONS

____ Prerequisite Skills Review: CRB pages 5–6
____ Strategies for Reading Mathematics: CRB pages 7–8
____ Homework Check: TE page 55: Answer Transparencies
____ Warm-Up or Daily Homework Quiz: TE pages 71 and 57, CRB page 11, or Transparencies

TEACHING OPTIONS

____ Lesson Opener (Visual Approach): CRB page 12 or Transparencies
____ Technology Activity with Keystrokes: CRB page 13
____ Examples 1–7: SE pages 71–74
____ Extra Examples: TE pages 72–74 or Transparencies; Internet
____ Closure Question: TE page 74
____ Guided Practice Exercises: SE page 75

APPLY/HOMEWORK

Homework Assignment

____ Basic 10–50 even, 55, 56, 64, 68, 74
____ Average 10–50 even, 51, 55, 56, 64, 68, 74
____ Advanced 10–50 even, 51, 54–60, 64, 68, 74

Reteaching the Lesson

____ Practice Masters: CRB pages 14–16 (Level A, Level B, Level C)
____ Reteaching with Practice: CRB pages 17–18 or Practice Workbook with Examples
____ Personal Student Tutor

Extending the Lesson

____ Applications (Interdisciplinary): CRB page 20
____ Challenge: SE page 78; CRB page 21 or Internet

ASSESSMENT OPTIONS

____ Checkpoint Exercises: TE pages 72–74 or Transparencies
____ Daily Homework Quiz (2.1): TE page 78, CRB page 24, or Transparencies
____ Standardized Test Practice: SE page 78; TE page 78; STP Workbook; Transparencies

Notes ____

____

____

LESSON 2.1

TEACHER'S NAME ________________ CLASS ______ ROOM ______ DATE ______

Lesson Plan for Block Scheduling

Half-day lesson (See *Pacing the Chapter,* TE pages 68C–68D) **For use with pages 71–78**

GOALS

1. **Recognize and analyze a conditional statement.**
2. **Write postulates about points, lines, and planes using conditional statements.**

State/Local Objectives ________________

____

____

CHAPTER PACING GUIDE	
Day	**Lesson**
1	Assess Ch. 1; **2.1 (all)**
2	2.2 (all)
3	2.3 (all)
4	2.4 (all); 2.5 (begin)
5	2.5 (end); 2.6 (begin)
6	2.6 (end); Review Ch. 2
7	Assess Ch. 2; 3.1 (all)

✓ Check the items you wish to use for this lesson.

STARTING OPTIONS

____ Prerequisite Skills Review: CRB pages 5–6
____ Strategies for Reading Mathematics: CRB pages 7–8
____ Homework Check: TE page 55: Answer Transparencies
____ Warm-Up or Daily Homework Quiz: TE pages 71 and 57, CRB page 11, or Transparencies

TEACHING OPTIONS

____ Lesson Opener (Visual Approach): CRB page 12 or Transparencies
____ Technology Activity with Keystrokes: CRB page 13
____ Examples 1–7: SE pages 71–74
____ Extra Examples: TE pages 72–74 or Transparencies; Internet
____ Closure Question: TE page 74
____ Guided Practice Exercises: SE page 75

APPLY/HOMEWORK

Homework Assignment

____ Block Schedule: 10–50 even, 51, 55, 56, 64, 68, 74

Reteaching the Lesson

____ Practice Masters: CRB pages 14–16 (Level A, Level B, Level C)
____ Reteaching with Practice: CRB pages 17–18 or Practice Workbook with Examples
____ Personal Student Tutor

Extending the Lesson

____ Applications (Interdisciplinary): CRB page 20
____ Challenge: SE page 78; CRB page 21 or Internet

ASSESSMENT OPTIONS

____ Checkpoint Exercises: TE pages 72–74 or Transparencies
____ Daily Homework Quiz (2.1): TE page 78, CRB page 24, or Transparencies
____ Standardized Test Practice: SE page 78; TE page 78; STP Workbook; Transparencies

Notes ________________

____

____

LESSON 2.1

NAME ______________________________ DATE __________

Available as a transparency

WARM-UP EXERCISES

For use before Lesson 2.1, pages 71–78

Use the diagram to name the points(s).

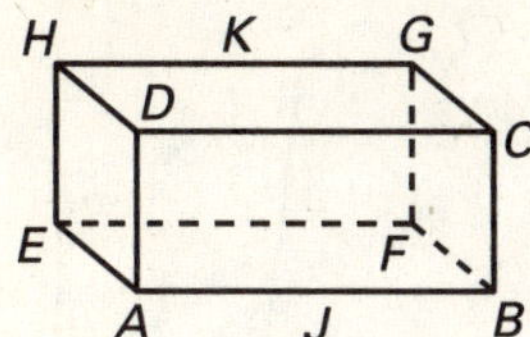

1. a point collinear with G and H

2. a point coplanar with D, C, and G

3. a point coplanar with H, G, and E

4. a point coplanar with A, D, E

DAILY HOMEWORK QUIZ

For use after Lesson 1.7, pages 51–58

Find the perimeter (or circumference) and area of each figure. Where necessary, use $\pi \approx 3.14$.

1. a circle with radius 4 units

2. a rectangle with length 8 units and width 2 units

3. 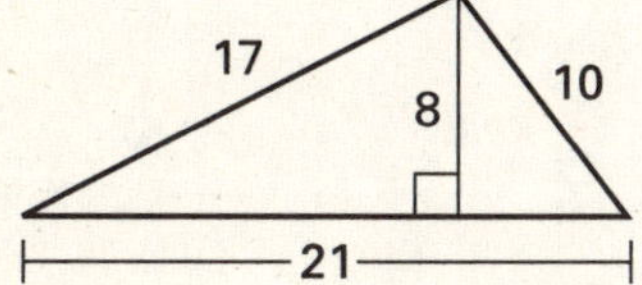

Find each unknown measurement.

4. the area of a square with perimeter 20 units

5. the circumference of a circle with area 452.2 m^2

Lesson 2.1

LESSON 2.1

NAME ______________________ DATE __________

Available as a transparency

Visual Approach Lesson Opener

For use with pages 71–78

Use the map at the right to determine whether the statement is *true* or *false*. Explain why your answer is correct.

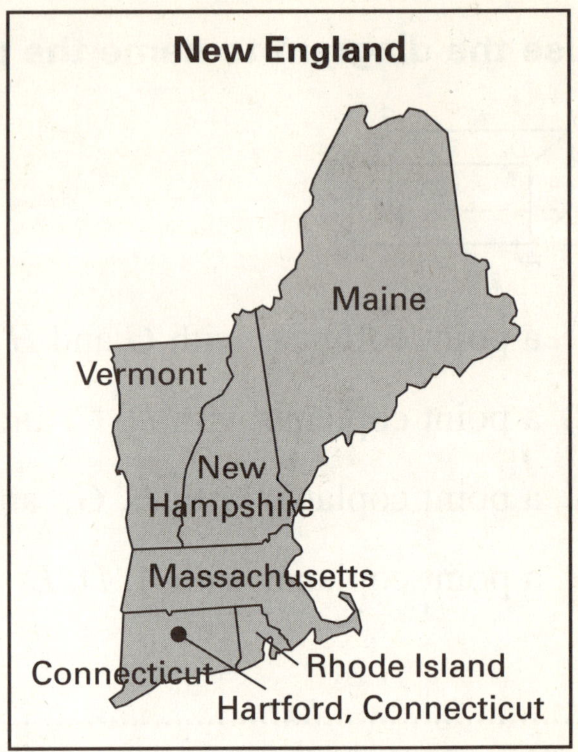

1. If you are in Massachusetts, then you are in New England.
2. If you are in New England, then you are in Massachusetts.
3. If you are in Hartford, then you are in Connecticut.
4. If you are in Connecticut, then you are in Hartford.

Use the Venn diagram at the right to determine whether the statement is *true* or *false*. Explain why your answer is correct.

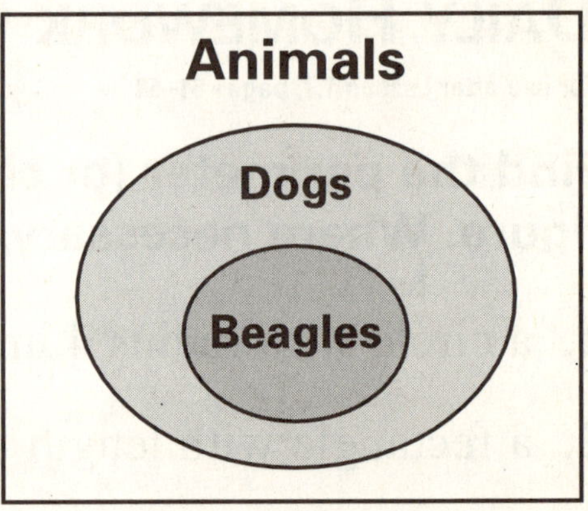

5. If an animal is a beagle, then it is a dog.
6. If an animal is a dog, then it is a beagle.
7. All animals are dogs.
8. All dogs are animals.
9. Draw a Venn diagram to show that all robins are birds, but not all birds are robins. Explain why your Venn diagram clearly shows this is true.

NAME ______________________ DATE __________

Technology Activity Keystrokes

For use with page 77

Keystrokes for Exercise 53

TI-92

1. Draw segment *AC*

   F2 5 (Locate cursor for point *A*) ENTER *A* (Move cursor to location for point *C*) ENTER *C*

2. Draw point *B* not on segment *AC*

   F2 1 (Locate desired position for point *B*) ENTER *B*

3. Draw segments *AB* and *BC*

   F2 5 (Locate cursor for point *A*) ENTER (Move cursor to point *B*) ENTER

   F2 5 (Locate cursor for point *B*) ENTER (Move cursor to point *C*) ENTER

4. Measure segments *AB*, *BC*, and *AC*

   F6 1 (Move cursor to segment *AB*) ENTER (Move cursor to segment *BC*) ENTER (Move cursor to segment *AC*) ENTER

5. Move *B* closer to *AC* and observe the measures of $\overline{AB}$, $\overline{BC}$, and $\overline{AC}$.

   F1 1 (Put cursor on point *B*) ENTER (Use the drag key [drag key] and the cursor pad to drag *B* closer to $\overline{AC}$)

SKETCHPAD

1. Draw segment *AC*. Choose the segment straightedge tool and draw segment *AC*. Select the text tool from the toolbox, double click on the label for point *B* and type *C*. Click OK.
2. Draw point *B* not on segment *AC*. Select the point tool and click on the screen. Relabel point *C* as point *B*.
3. Construct segments *AB* and *BC*. Choose the segment straightedge tool and click and hold on point *A* and move mouse to point *B*. Repeat this process to draw segment *BC*.
4. Measure segments *AB*, *BC*, and *AC*. Choose the selection arrow tool. Select segments *AB, BC,* and *AC* (hold down the shift key). Choose **Length** from the **Measure** menu.
5. Move *B* closer to *AC* and observe the measures of $\overline{AB}$, $\overline{BC}$, and $\overline{AC}$. Drag *B* by clicking it and holding down the mouse button (make sure the selection arrow tool is in translate mode).

LESSON 2.1

NAME ______________________________ DATE ____________

Practice A

For use with pages 71–78

Identify the hypothesis and the conclusion.

1. If the weather is warm, then we should go swimming.
2. If you want good service, then take your car to Joe's Service Center.
3. If you like purple, you'll love this sweater.
4. $2x - 12 = 40$ only if $x = 26$.
5. If the groundhog sees its shadow, then there will be six more weeks of winter.

Rewrite the conditional statement in if-then form.

6. Today is Monday if yesterday was Sunday.
7. An object measures 12 inches if it is one foot long.
8. A number is divisible by 4 if it is divisible by 8.
9. An acute angle is an angle that measures less than 90°.
10. Geometry is offered only during periods 2 and 4.

Decide whether the statement is *true* or *false*. If false, provide a counterexample.

11. The equation $2x - 7 = 5 + x$ has exactly one solution.
12. If $x^2 = 16$, then x must equal 8 or -8.
13. February 14 is Valentine's Day.
14. If you visited the Statue of Liberty, then you've been to New York.
15. A point may lie on at most two lines.

Write the converse and contrapositive of each statement.

16. If you like tennis, then you play on the tennis team.
17. If x is odd, then $2x$ is even.
18. If $m\angle P = 45°$, then $\angle P$ is acute.

Use the diagram to state the postulate(s) that verifies the truth of the statement.

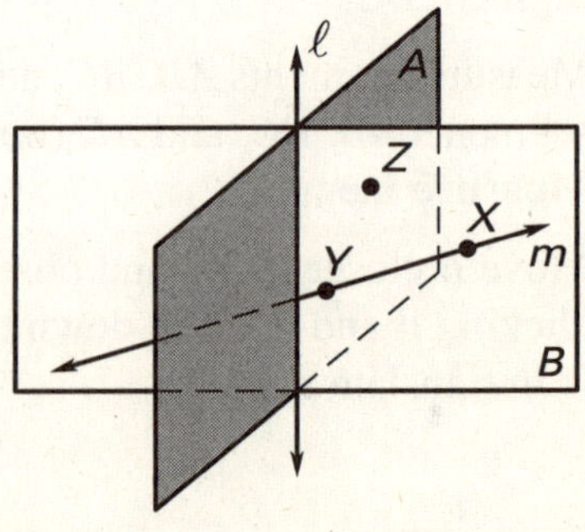

19. The points X, Y, and Z lie in a plane (labeled B).
20. The points X and Y lie on a line (labeled m).
21. The planes A and B intersect in a line (labeled l).
22. The points X and Y lie in a plane B. Therefore, line m lies in plane B.

LESSON 2.1

NAME ______________________________ DATE __________

Practice B

For use with pages 71–78

Rewrite the conditional statement in if-then form.

1. It is time for dinner if it is 6 P.M.
2. There are 12 eggs if the carton is full.
3. A number is divisible by 6 if it is divisible by 2 and 3.
4. An obtuse angle is an angle that measures more than 90° and less than 180°.
5. Geometry is offered only during even numbered blocks.

Decide whether the statement is *true* or *false*. If false, provide a counterexample.

6. The equation $4x - 3 = 12 + 2x$ has exactly one solution.
7. If $x^2 = 36$, then x must equal 18 or -18.
8. Thanksgiving is celebrated on a Thursday.
9. If you visited Springfield, then you've been to Illinois.
10. Two lines intersect in at most one point.

Write the converse, inverse, and contrapositive of each statement.

11. If you like hockey, then you go to the hockey game.
12. If x is odd, then $3x$ is odd.
13. If $m\angle P = 90°$, then $\angle P$ is a right angle.

Draw a sketch to illustrate each postulate.

14. If two lines intersect, then their intersection is exactly one point.
15. If two points lie in a plane, then the line containing them lies in the plane.
16. If two planes intersect, then their intersection is a line.

Use the diagram to state the postulate(s) that verifies the truth of the statement.

17. The points E, F, and H lie in a plane (labeled R).
18. The points E and F lie on a line (labeled m).
19. The planes Q and R intersect in a line (labeled l).
20. The points E and F lie in a plane R. Therefore, line m lies in plane R.

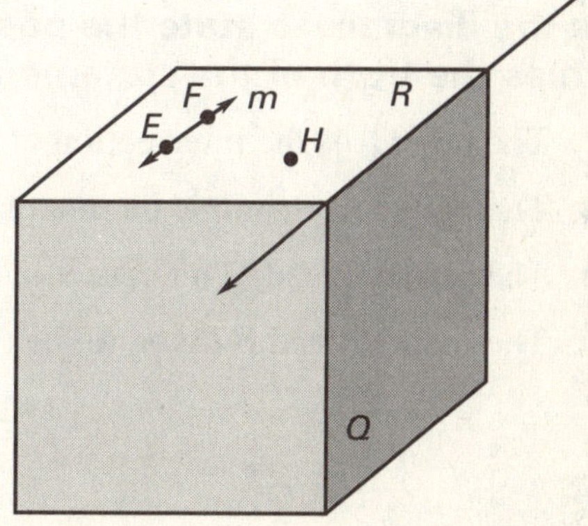

Lesson 2.1

LESSON

2.1

NAME ______________________________ DATE ____________

Practice C

For use with pages 71–78

Rewrite the conditional statement in if-then form.

1. I will go to the game if I get all of my homework done.
2. The water will freeze if the temperature is 10°F.
3. A student on the high honor roll has at least a 90 average.
4. Bert goes shopping for groceries only on Wednesday.
5. The number 2 is a factor of every even number.

Decide whether the statement is *true* or *false*. If false, provide a counterexample.

6. The equation $-3x - 10 = 5 + 2x$ has exactly one solution.
7. If $x > 0$, then $x^2 > x$.
8. For any real numbers a and b, $|a + b| = |a| + |b|$.
9. If you visited the Jeffereson Monument, then you've been to Washington, D.C.
10. Two collinear rays intersect.

Write the converse, inverse, and contrapositive of each statement. Identify each statment at *true* or *false*.

11. If you like volleyball, then you like to be at the beach.
12. If $4x$ is even, then x is odd.
13. If $m\angle P = 109°$, then $\angle P$ is obtuse.

Draw a sketch to illustrate each postulate.

14. A line contains at least two points.
15. Through any three noncollinear points there exists exactly one plane.
16. A plane contains at least three noncollinear points.

Use the diagram to state the postulate(s) that verifies the truth of the statement.

17. The point A is the intersection of lines l and m.
18. The points A, B, and C lie in a plane (labeled Q).
19. The planes P and Q interesect in a line (labeled l).
20. The points A and B lie on a line (labeled m).

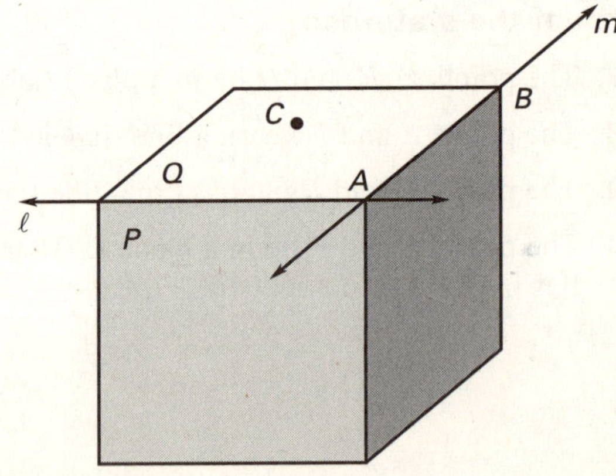

LESSON 2.1

NAME ______________________ DATE ________

Reteaching with Practice

For use with pages 71–78

GOAL **Recognize and analyze a conditional statement and write postulates about points, lines, and planes using conditional statements**

Vocabulary

A **conditional statement** has two parts, a *hypothesis* and a *conclusion.*

When the statement is written in **if-then form,** the "if" part contains the **hypothesis** and the "then" part contains the **conclusion.**

The **converse** of a conditional statement is formed by switching the hypothesis and conclusion.

A statement can be altered by **negation,** that is, by writing the negative of the statement.

When you negate the hypothesis and conclusion of a conditional statement, you form the **inverse.**

When you negate the hypothesis and conclusion of the converse of a conditional statement, you form the **contrapositive.**

When two statements are both true or both false, they are called **equivalent statements.**

Postulate 5 Through any two points there exists exactly one line.

Postulate 6 A line contains at least two points.

Postulate 7 If two lines intersect, then their intersection is exactly one point.

Postulate 8 Through any three noncollinear points there exists exactly one plane.

Postulate 9 A plane contains at least three noncollinear points.

Postulate 10 If two points lie in a plane, then the line containing them lies in the plane.

Postulate 11 If two planes intersect, then their intersection is a line.

EXAMPLE 1 ***Writing the If-Then Form, Inverse, Converse, and Contrapositive***

a. Rewrite the conditional statement in *if-then* form.

b. Write the inverse, converse, and contrapositive of the statement.
An even number greater than two is not prime.

SOLUTION

a. If a number greater than two is even, then it is not prime.

b. **inverse:** If a number greater than two is not even, then it is prime.
converse: If a number greater than two is not prime, then it is even.
contrapositive: If a number greater than two is prime, then it is not even.

Lesson 2.1

LESSON 2.1 CONTINUED

NAME ______________________ DATE ____________

Reteaching with Practice

For use with pages 71–78

Exercises for Example 1

For the conditional statement (a) rewrite it in *if-then* form, (b) write the inverse, (c) write the converse, and (d) write the contrapositive.

1. I will dry the dishes if you will wash them.
2. A square with side of length 3 centimeters has an area of 9 square centimeters.
3. An angle with measure of 90° is a right angle.

EXAMPLE 2 Using Postulates and Counterexamples

Decide whether the statement is *true* or *false.* If it is false, give a counterexample.

a. Through any three points there exists exactly one line.

b. For any one point in a plane, there exists exactly one line through that point in that plane.

Solution

a. This statement is false. If the three points are noncollinear, then there is no line which includes all three points.

b. This statement is false. In the diagram at the right, lines *l* and *m* both go through point *A* and both are in plane *P.*

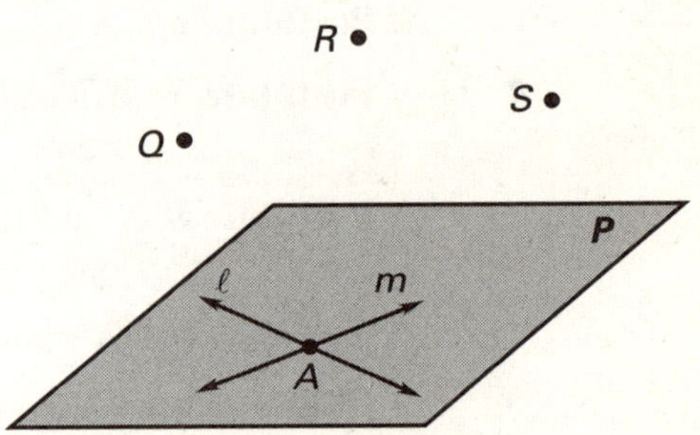

Exercises for Example 2

Decide whether the statement is true or false. If it is false, give a counterexample.

4. Three planes can intersect in a point.
5. For noncollinear points *A, B,* and *C,* point *C* and the line formed by *A* and *B* are in different planes.

LESSON 2.1

NAME ____________________ DATE ________

Quick Catch-Up for Absent Students

For use with pages 71–78

The items checked below were covered in class on (date missed) ________

Lesson 2.1: Conditional Statements

____ **Goal 1:** Recognize and analyze a conditional statement. (pp. 71–72)

Material Covered:

____ Example 1: Rewriting in If-Then Form

____ Example 2: Writing a Counterexample

____ Example 3: Writing the Converse of a Conditional Statement

____ Example 4: Writing an Inverse, Converse, and Contrapositive

Vocabulary:

conditional statement, p. 71
hypothesis, p. 71
converse, p. 72
inverse, p. 72
equivalent statements, p. 72
if-then form, p. 71
conclusion, p. 71
negation, p. 72
contrapositive, p.72

____ **Goal 2:** Write postulates about points, lines, and planes using conditional statements. (pp. 73–74)

Material Covered:

____ Student Help: Study Tip

____ Example 5: Identifying Postulates

____ Example 6: Rewriting a Postulate

____ Student Help: Study Tip

____ Example 7: Using Postulates and Counterexamples

____ Other (specify) ____________________

____

Homework and Additional Learning Support

____ Textbook (specify) pp. 75–78

____ Internet: Extra Examples at www.mcdougallittell.com

____ *Reteaching with Practice* worksheet (specify exercises) ________

____ *Personal Student Tutor* for Lesson 2.1

Lesson 2.1

LESSON
2.1

NAME ______________________________ DATE __________

Interdisciplinary Application

For use with pages 71–78

Conditional Statement

LITERATURE

Read the following passage from *Alice's Adventures in Wonderland* by Lewis Carroll and then answer the questions.

"Then you should say what you mean," the March Hare went on.

"I do," Alice hastily replied; "at least—at least I mean what I say—that's the same thing, you know."

"Not the same thing a bit!" said the Hatter. "You might just as well say that 'I see what I eat' is the same thing as 'I eat what I see'!". . .

"You might just as well say," added the Dormouse, who seemed to be talking in its sleep, "that 'I breathe when I sleep' is the same thing as 'I sleep when I breathe'!"

"It IS the same thing with you," said the Hatter, and here the conversation dropped. . . .

1. How many conditional statements are in the passage?
2. How many inverses are there?
3. How many converses are there?
4. How many contrapositives are listed?

LESSON 2.1

NAME ______________________________ DATE __________

Challenge: Skills and Applications

For use with pages 71–78

Find the inverse, converse, and contrapositive of the statement. (You may need to rewrite the statement in if-then form first.)

1. If a bicycle is red, then it has two seats.
2. If a thopfoo can make a pucho, then all gorups are purple.
3. A student who enjoys mathematics certainly must have a logical mind.
4. All citizens of the Regal Kingdom have blue hair.

Tell whether the statement is *true* or *false*, state the converse of the statement, and tell whether the converse is *true* or *false*. If either is false, give a counterexample.

5. If a positive integer is divisible by 15, then it is divisible by both 3 and 5.
6. If a positive integer is divisible by both 2 and 3, then it is divisible by 12.
7. For an integer n, if n^2 is divisible by 3, then n is divisible by 3.
8. For an integer n, if n^2 is divisible by 4, then n is divisible by 4.

Tell whether the statement is *true* or *false*. Explain your answer.

9. The intersection of any two planes is a line.
10. Through any two points there exists exactly one plane.
11. A plane contains of at least three points.
12. If three points lie in a plane, the points are noncollinear.

In Exercises 13–18, let *a* and *b* be two intersecting lines that intersect at point *X*.

13. Make a conjecture about the number of planes that contain both a and b.
14. Which postulate allows you to state that there is a point Y, distinct from X, on line a, and a point Z, distinct from X, on line b?
15. Which postulate guarantees that point Y is not on line b?
16. Which postulate allows you to conclude that there is exactly one plane P that contains points X, Y, and Z?
17. Which postulate guarantees that lines a and b are contained in plane P?
18. Was the conjecture you made in Exercise 13 correct?

LESSON 2.2

TEACHER'S NAME ______________ CLASS ______ ROOM ______ DATE ______

Lesson Plan

2-day lesson (See *Pacing the Chapter,* TE pages 68C–68D)

For use with pages 79–85

GOAL

1. **Recognize and use definitions.**
2. **Recognize and use biconditional statements.**

State/Local Objectives ______________________________

____

✓ Check the items you wish to use for this lesson.

STARTING OPTIONS

____ Homework Check: TE page 75: Answer Transparencies
____ Warm-Up or Daily Homework Quiz: TE pages 79 and 78, CRB page 24, or Transparencies

TEACHING OPTIONS

____ Motivating the Lesson: TE page 80
____ Lesson Opener (Geometry Software): CRB page 25 or Transparencies
____ Technology Activity with Keystrokes: CRB pages 26–29
____ Examples: Day 1: 1–3, SE pages 79–80; Day 2: 4–5, SE page 81
____ Extra Examples: Day 1: TE page 80 or Transp.; Day 2: TE page 81 or Transp.; Internet
____ Closure Question: TE page 81
____ Guided Practice Exercises: SE page 82   Day 1: Exs. 1–8; Day 2: Exs. 9–12

APPLY/HOMEWORK

Homework Assignment

____ Basic   Day 1: 13–31; Day 2: 32–44, 46, 50–55, 60-68
____ Average   Day 1: 13–31; Day 2: 32–44, 46, 50–55, 60-68
____ Advanced   Day 1: 13–31; Day 2: 32–44, 46, 50–58, 60-68

Reteaching the Lesson

____ Practice Masters: CRB pages 30–32 (Level A, Level B, Level C)
____ Reteaching with Practice: CRB pages 33–34 or Practice Workbook with Examples
____ Personal Student Tutor

Extending the Lesson

____ Applications (Real-Life): CRB page 36
____ Challenge: SE page 85; CRB page 37 or Internet

ASSESSMENT OPTIONS

____ Checkpoint Exercises: Day 1: TE page 80 or Transp.; Day 2: TE page 81 or Transp.
____ Daily Homework Quiz (2.2): TE page 85, CRB page 40, or Transparencies
____ Standardized Test Practice: SE page 85; TE page 85; STP Workbook; Transparencies

Notes ______________________________

____

____

LESSON 2.2

TEACHER'S NAME ______ CLASS ______ ROOM ______ DATE ______

Lesson Plan for Block Scheduling

1-day lesson (See *Pacing the Chapter,* TE pages 68C–68D) For use with pages 79–85

GOALS
1. **Recognize and use definitions.**
2. **Recognize and use biconditional statements.**

State/Local Objectives ______

CHAPTER PACING GUIDE	
Day	**Lesson**
1	Assess Ch. 1; 2.1 (all)
2	**2.2 (all)**
3	2.3 (all)
4	2.4 (all); 2.5 (begin)
5	2.5 (end); 2.6 (begin)
6	2.6 (end); Review Ch. 2
7	Assess Ch. 2; 3.1 (all)

✓ Check the items you wish to use for this lesson.

STARTING OPTIONS

____ Homework Check: TE page 75: Answer Transparencies
____ Warm-Up or Daily Homework Quiz: TE pages 79 and 78, CRB page 24, or Transparencies

TEACHING OPTIONS

____ Motivating the Lesson: TE page 80
____ Lesson Opener (Geometry Software): CRB page 25 or Transparencies
____ Technology Activity with Keystrokes: CRB pages 26–29
____ Examples: 1–5, SE pages 79–81
____ Extra Examples: TE pages 80–81 or Transparencies; Internet
____ Closure Question: TE page 81
____ Guided Practice Exercises: SE page 82

APPLY/HOMEWORK

Homework Assignment

____ Block Schedule: 13–44, 46, 50–55, 60-68

Reteaching the Lesson

____ Practice Masters: CRB pages 30–32 (Level A, Level B, Level C)
____ Reteaching with Practice: CRB pages 33–34 or Practice Workbook with Examples
____ Personal Student Tutor

Extending the Lesson

____ Applications (Real-Life): CRB page 36
____ Challenge: SE page 85; CRB page 37 or Internet

ASSESSMENT OPTIONS

____ Checkpoint Exercises: TE pages 80–81 or Transparencies
____ Daily Homework Quiz (2.2): TE page 85, CRB page 40, or Transparencies
____ Standardized Test Practice: SE page 85; TE page 85; STP Workbook; Transparencies

Notes ______

NAME ______________________ DATE ________

Available as a transparency

Warm-Up Exercises

For use before Lesson 2.2, pages 79–85

Write the converse of each statement.

1. If M is the midpoint of $\overline{AB}$, then $AM = MB$.
2. If $AB + BC = AC$, then points A, B, and C are collinear.
3. If $|x| = 6$, then $x = 6$ or -6.
4. You score a goal if the ball crosses the goal line.

Daily Homework Quiz

For use after Lesson 2.1, pages 71–78

Using the true conditional statement, "Constrictors such as pythons are not poisonous," write each of the following.

1. Rewrite the statement in if-then form.
2. Write the converse of the if-then statement.
3. Write the contrapositive of the if-then statement.
4. Write the hypothesis and conclusion of the if-then statement.

LESSON 2.2

NAME ________________ DATE ________

Available as a transparency

Geometry Software Lesson Opener

For use with pages 79–85

Use the statement below to answer Questions 1–4.

If a triangle has one right angle, then the triangle has two complementary angles.

1. Do you think the statement is true? Explain.
2. Use geometry software to draw several triangles with one right angle. Measure the other two angles in each triangle. How do your results verify your answer to Question 1?
3. Write the converse of the statement. Do you think the converse is true? Explain.
4. Use geometry software to draw several triangles with two complementary angles. Measure the third angle in each triangle. How do your results verify your answer to Question 3?

Use the statement below to answer Questions 5–8. Recall that a polygon is a plane figure formed by three or more segments, called sides, such that no two sides with a common endpoint are collinear, and each side intersects exactly two other sides, one at each endpoint.

If a polygon has exactly four angles and each is a right angle, then the polygon is a rectangle.

5. Do you think this statement is true? Explain.
6. Use geometry software to draw several polygons with exactly four angles, each measuring 90°. What is true about all of the polygons you have drawn? How do the results verify your answer to Question 5?
7. Write the converse of the statement. Do you think the converse is true? Explain.
8. Use geometry software to draw several rectangles. Measure the four angles of each rectangle. How do the results verify your answer to Question 7?

LESSON 2.2

NAME ______________________ DATE __________

Technology Activity

For use with pages 79–85

GOALS **To determine if a biconditional statement is true or false by testing both the statement and its converse**

Show that the following biconditional statement is true by showing that the conditional statement and its converse are true.

Biconditional statement:
The three sides of a triangle are congruent if and only if the three angles of the triangle are congruent.

Activity

1. Show that the conditional statement is true. Construct a side of the triangle.
2. Use the compass feature of the software to construct the other two sides of the triangle.
3. Measure each side of the triangle.
4. Measure each angle of the triangle.
5. Grab an endpoint of the original side and drag the point.

Exercises

1. Rewrite the biconditional statement as a conditional statement and its converse.
2. *Extension* Show that the converse of the conditional statement is true using the following steps.
   a. Construct a circle with a radius of AB and centered at point A.
   b. Construct a second circle using point B as the center and using the same radius as the first circle.
   c. Choose one of the intersection points of the two circles as the third vertex of the triangle along with A and B. This is a construction to create $60°$ angles.
   d. Draw the remaining two segments of the triangle.
   e. Measure each angle.
   f. Measure each side.

NAME ______________________ DATE ____________

Technology Activity Keystrokes

For use with pages 79–85

TI-92

1. Construct segment *AB*

   F2 5 (Place cursor at desired location for point *A*) ENTER *A* (Move cursor to location for *B*) ENTER *B*

2. Use the compass feature to locate vertex *C*

   F4 8 (Place cursor on point *A*) ENTER (Place cursor on segment *AB*) ENTER

   F4 8 (Place cursor on point *B*) ENTER (Place cursor on segment *AB*) ENTER

   Label one of the intersections of the two circles as point *C*

   F2 3 (Place cursor at the intersection of the two circles) ENTER *C*

   Hide the two circles

   F7 1 (Place cursor on the first circle) ENTER (Move cursor to second circle) ENTER

   Construct segments *AC* and *BC*

   F2 5 (Place cursor on point *A*) ENTER (Move cursor to point *C*) ENTER

   F2 5 (Place cursor on point *B*) ENTER (Move cursor to point *C*) ENTER

3. Measure the lengths of sides *AB*, *BC*, and *CA*

   F6 1 (Place cursor on segment *AB*) ENTER (Move cursor to segment *BC*) ENTER

   (Move cursor to segment *CA*) ENTER

4. Measure angles *ABC*, *BCA*, and *CAB*

   F6 3 (Place cursor on point *A*) ENTER (Move cursor to point *B*) ENTER (Move cursor to point *C*) ENTER

   Repeat for remaining angles

5. Grab point *A* and drag it

   F1 1 (Place cursor on point *A*) ENTER (Use drag key and the cursor pad to drag the point)

NAME ______________________________ DATE __________

Technology Activity Keystrokes

For use with pages 79–85

SKETCHPAD

1. Construct segment *AB*. Choose the segment straightedge tool.

2. Use compass to locate vertex *C*. Choose compass tool from the toolbox. Construct a circle with center point *A* and radius point *B*. Use the Tool Status Box in the lower left-hand corner to ensure that point *A* and point *B* are being used. Repeat with point *B* as the center and point *A* as the radius point. Choose the Pointer tool, and plot point *C*, on the intersection of the two circles.

3. Measure the lengths of *AC*, *BC*, and *AB*. Choose the Selection arrow and then select each segment while holding down the shift key. Select **Length** from the **Measure** menu.

4. Measure angles *ABC*, *BCA*, and *CAB*. To measure angle *ABC*, select points *A*, *B*, and *C*. Then select **Angle** from the **Measure** menu. Repeat for remaining angles. Before selecting the next angle, be sure to click anywhere in the work area to deselect the previous points.

5. Grab point *A* and drag it.

LESSON 2.2

NAME ______________________ DATE ________

Practice A

For use with pages 79–85

Use the diagram to determine whether the statement is *true* or *false*.

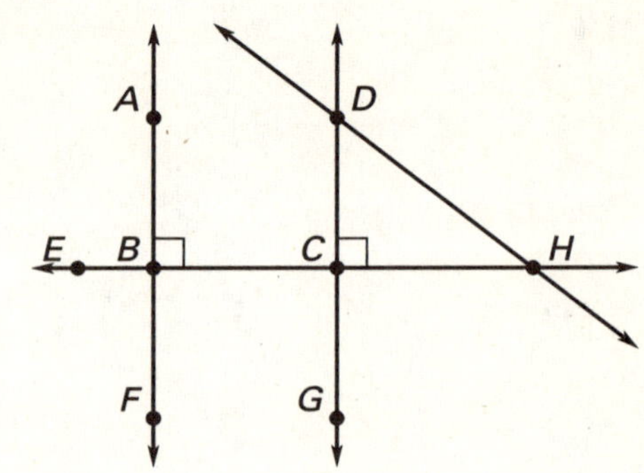

1. Points A, B, and C are collinear.
2. $\angle DCB$ and $\angle DCH$ are supplementary.
3. Points E, D, and H lie in the same plane.
4. $\overleftrightarrow{DH}$ is perpendicular to $\overleftrightarrow{EH}$.
5. $\overleftrightarrow{HE}$ is perpendicular to $\overleftrightarrow{AF}$.
6. $\angle DCB$ and $\angle ABC$ are complementary.
7. Point C is the midpoint of $\overline{BH}$.

Rewrite the biconditional statement as a conditional statement and its converse.

8. Two segments are congruent if and only if they have the same measure.
9. Three points are collinear if and only if they lie on the same line.
10. Four points are coplanar if and only if they lie in the same plane.
11. You may go to the movies Friday night if and only if you clean your room.
12. You may become president of the United States if and only if you are 35 years old.

Lesson 2.2

Give a counterexample that demonstrates that the converse of the statement is false.

13. If you live in Detroit, then you live in Michigan.
14. If an angle measures 30°, then it is acute.
15. If an animal is a leopard, then it has spots.
16. If the month is September, then there are 30 days in the month.
17. If two angles are vertical angles, then they are not adjacent.

In Exercises 18 and 19, use the information in the table to write a definition for each type of saxophone. The first one is started for you.

	Frequency (cycles per second)	
Instrument	***Lower limit***	***Upper limit***
E-flat baritone saxophone	69	416
B-flat tenor saxophone	104	622
E-flat alto saxophone	138	831

Sample: A saxophone that has a frequency of 69 cycles per second to 416 cycles per second is called an E-flat baritone saxophone.

18. B-flat tenor saxophone
19. E-flat alto saxophone

LESSON **2.2**

NAME ______________________________ DATE __________

Practice B

For use with pages 79–85

Use the diagram to determine whether the statement is *true* or *false*.

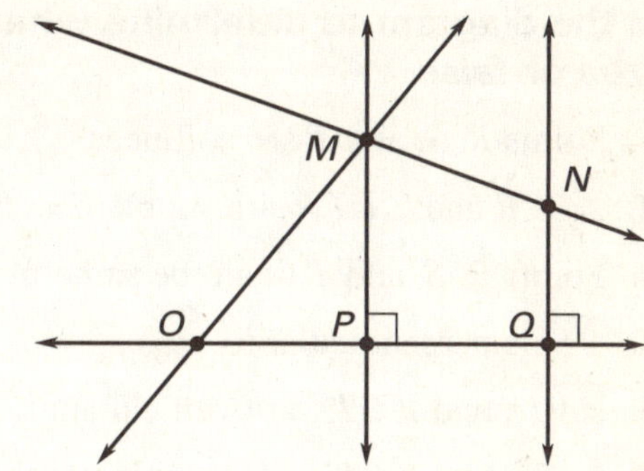

1. Points O, P, and Q are collinear.
2. $\angle MPO$ and $\angle NQP$ are supplementary.
3. Points M, P, and O lie in the same plane.
4. $\overleftrightarrow{MP}$ is perpendicular to $\overleftrightarrow{NQ}$.
5. $\overleftrightarrow{NQ}$ is perpendicular to $\overleftrightarrow{OQ}$.
6. $\angle MPO$ and $\angle MPQ$ are complimentary.
7. Point P is the midpoint of $\overline{OQ}$.

Rewrite the biconditional statement as a conditional statement and its converse.

8. $x = 4$ if and only if $x^2 = 16$.
9. Point Y lies between points X and Z if and only if $XY + YZ = XZ$.
10. The car will run if and only if there is gas in the tank.
11. Two angles are congruent if and only if they have the same measure.
12. An angle is a right angle if and only if it measures 90°.

Write the converse of each true statement. If the converse is also true, combine the statements to write a true biconditional statement.

13. If you live in Detroit, then you live in Michigan.
14. If an angle measures 30°, then it is acute.
15. If two angles are supplementary, then their sum is 180°.
16. If two angles are congruent, then they have the same measure.
17. If two angles are vertical angles, then they are not adjacent.

In Exercises 18–20, use the information in the table to write a definition for each type of saxophone.

	Frequency (cycles per second)	
Instrument	*Lower limit*	*Upper limit*
E-flat baritone saxophone	69	416
B-flat tenor saxophone	104	622
E-flat alto saxophone	138	831

18. E-flat baritone saxophone
19. B-flat tenor saxophone
20. E-flat alto saxophone

LESSON **2.2**

NAME ______________________ DATE __________

Practice C

For use with pages 79–85

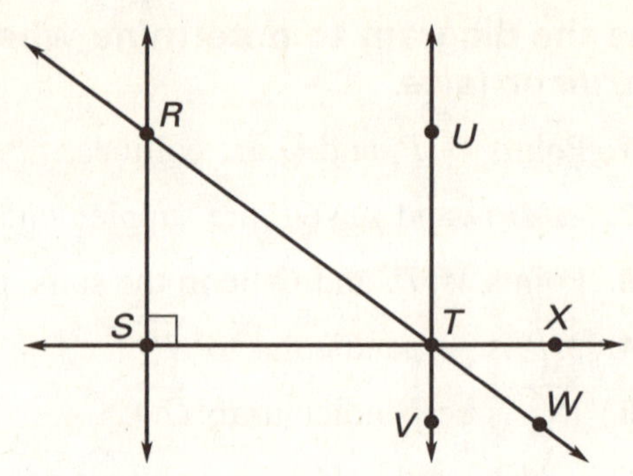

Use the diagram to determine whether the statement is *true* or *false*.

1. Points R, S, and T are collinear.
2. $\angle UTR$ and $\angle UTW$ are supplementary.
3. Points R, S, and T lie in the same plane.
4. $\overleftrightarrow{TS}$ is perpendicular to $\overleftrightarrow{RS}$.
5. $\angle VTS$ and $\angle UTX$ are vertical angles.
6. $\angle STR$ and $\angle RTU$ are complementary.
7. Line $\overleftrightarrow{RW}$ bisects $\angle UTS$.

Rewrite the biconditional statement as a conditional statement and its converse.

8. An angle is acute if and only if it measures less than 90°.
9. Three points are collinear if and only if they lie on the same line.
10. I eat pizza if and only if it is Friday night.
11. The game is cancelled if and only if it rains.
12. A number is divisible by 6 if and only if it is divisible by 2 and 3.

Lesson 2.2

Write the converse of each true statement. If the converse is also true, combine the statements to write a true biconditional statement. If the converse is false, give a counterexample.

13. If you live in Detroit, then you live in Michigan.
14. If an angle measures 30°, then it is acute.
15. If two angles are supplementary, then their sum is 180°.
16. If two angles are congruent, then they have the same measure.
17. If two angles are vertical angles, then they are not adjacent.

In Exercises 18–20, use the information in the table to write a definition for each type of saxophone.

Instrument	*Frequency (cycles per second)*	
	Lower limit	*Upper limit*
E-flat baritone saxophone	69	416
B-flat tenor saxophone	104	622
E-flat alto saxophone	138	831

18. E-flat baritone saxophone
19. B-flat tenor saxophone
20. E-flat alto saxophone
21. If the frequency of a saxophone was 100 Hz, what could you conclude?
22. If the frequency of a saxophone was 150 Hz, what could you conclude?

LESSON 2.2

NAME ______________________ DATE __________

Reteaching with Practice

For use with pages 79–85

GOAL Recognize and use biconditional statements

VOCABULARY

A **biconditional statement** is a statement that contains the phrase "if and only if."

EXAMPLE 1 *Rewriting a Biconditional Statement*

Rewrite the biconditional statement as a conditional statement and its converse.

a. An angle is a right angle if and only if its measure is 90°.

b. A number is even if and only if it is divisible by two.

c. A point on a segment is the midpoint of the segment if and only if it bisects the segment.

SOLUTION

a. Conditional statement: If an angle is a right angle, then its measure is 90°.

Converse: If an angle's measure is 90°, then it is a right angle.

b. Conditional statement: If a number is even, then it is divisible by two.

Converse: If a number is divisible by two, then it is even.

c. Conditional statement: If a point on a segment is the midpoint of the segment, then it bisects the segment.

Converse: If a point on a segment bisects the segment, then it is the midpoint of the segment.

Exercises for Example 1

Rewrite the biconditional statement as a conditional statement and its converse.

1. A number is a perfect square if and only if it is the product of some number times itself.
2. Two angles are complementary if and only if their sum is 90°.
3. A real number is rational if and only if it can be written in the form $\frac{p}{q}$, where p and q are integers and $q \neq 0$.

EXAMPLE 2 *Analyzing a Biconditional Statement*

Consider the following statement: You attend school if and only if it is a weekday.

a. Is this a biconditional statement?

b. Is the statement true?

Lesson 2.2

NAME ______________________ DATE ____________

Reteaching with Practice

For use with pages 79–85

SOLUTION

a. The statement is a biconditional statement because it contains the words "if and only if."

b. The statement can be rewritten as the following statement and its converse.

Conditional statement: If you are in school, then it is a weekday.

Converse: If it is a weekday, then you are in school.

The first of these statements is true, but the second is false (consider weekdays during the summer, or holidays that occur on a weekday). So, the biconditional statement is false.

Exercises for Example 2

Consider the given statement and decide (a) if it is a biconditional statement and (b) if the statement is true.

4. It rains if and only if there are clouds in the sky.

5. Two angles are adjacent angles if and only if they share a vertex and one side but do not have any common interior points.

EXAMPLE 3 *Writing a Biconditional Statement*

The following statement is true. Write the converse of the statement and decide whether the converse is *true* or *false.* If the converse is true, combine it with the original statement to form a true biconditional statement. If the converse is false, state a counterexample.

If the product ab is negative, then either a is negative or b is negative.

SOLUTION

Converse: If either a is negative or b is negative, then the product ab is negative.

The converse is false. Consider the counterexample where $a = -3$ but $b = 0$. Then the product ab would be 0, which is not negative.

Exercises for Example 3

Each of the following statements is true. Write the converse of the statement and decide whether the converse is *true* or *false.* If the converse is true, combine it with the original statement to form a true biconditional statement. If the converse is false, state a counterexample.

6. If the sides of two angles form two pairs of opposite rays, then the angles are vertical angles.

7. If the product ab is 0, then either a must be 0 or b must be zero.

LESSON
2.2

NAME ______________________________ DATE __________

Quick Catch-Up for Absent Students

For use with pages 79–85

The items checked below were covered in class on (date missed) __________

Lesson 2.2: Definitions and Biconditional Statements

____ **Goal 1:** Recognize and use definitions. (p. 79)

Material Covered:

____ Example 1: Using Definitions

Vocabulary:

perpendicular lines, p. 79 line perpendicular to a plane, p. 79

____ **Goal 2:** Recognize and use biconditional statements. (pp. 80–81)

Material Covered:

____ Student Help: Study Tip

____ Example 2: Rewriting a Biconditional Statement

____ Example 3: Analyzing a Biconditional Statement

____ Example 4: Writing a Biconditional Statement

____ Student Help: Study Tip

____ Example 5: Writing a Postulate as a Biconditional

Vocabulary:

biconditional statement, p. 80

____ Other (specify) ______________________________

____

Homework and Additional Learning Support

____ Textbook (specify) pp. 82–85

____

____ *Reteaching with Practice* worksheet (specify exercises) __________

____ *Personal Student Tutor* for Lesson 2.2

Lesson 2.2

NAME ______________________ DATE ________

Real-Life Application: When Will I Ever Use This?

For use with pages 79–85

Discipline Codes

In your high school handbook, a section is devoted to a discipline code that students are to abide by. Rules and consequences for violating the code are well defined. In most school districts, students are given copies of the code, discuss it during homeroom, and sign a statement that they have *received* the code.

The following are conditional statements from a sample discipline code. Write the converse of each statement. Determine if the conditional statement and the converse are biconditional.

1. If you are late to any class, you will be assigned an office detention.
2. If you are insubordinate to any adult in the building, you will be assigned a Saturday detention.
3. In any computer environment, if you are using unauthorized access to the internet, you will lose computer privileges for one week in all of your classes.
4. If you are involved in a verbal argument with another *student*, you and that student will be required to attend conflict resolution sessions.
5. If you skip a day of school, you will serve three days of ISS (in-school-suspension).

LESSON 2.2

NAME ______________________ DATE __________

Challenge: Skills and Applications

For use with pages 79–85

A *truth table* can be used to show how the truth or falsity of a statement depends on the truth or falsity of its parts. Assume that p and q represent statements that can be either true (T) or false (F).

We will need to introduce some notation:

$\sim p$ means "not p"

$p \rightarrow q$ means "p implies q," or equivalently, "if p, then q."

$p \leftrightarrow q$ means "if p, then q and if q, then p"

Shown below is the truth table for $\sim p$. It lists both possible truth values for p and the corresponding truth values for $\sim p$.

p	$\sim p$
T	F
F	T

Shown below is the truth table for $p \rightarrow q$. Because the conditional statement $p \rightarrow q$ is based on two statements, p and q, there are four possible situations: both p and q are true, p is true but q is false, p is false but q is true, or both p and q are false.

p	q	$p \rightarrow q$
T	T	T
T	F	F
F	T	T
F	F	T

It is generally accepted that when the hypothesis of a conditional statement is false, the conditional statement is true.

1. Construct a truth table for the converse of $p \rightarrow q$.

2. Construct a truth table for the inverse of $p \rightarrow q$ using the table headings shown below.

p	q	$\sim p$	$\sim q$	$\sim p \rightarrow \sim q$

3. Construct a truth table for the contrapositive of $p \rightarrow q$.

4. Construct a truth table for the biconditional statement $p \rightarrow q$.

5. a. Construct a truth table for the statement $\sim p \leftrightarrow \sim q$.

b. What simpler statement is equivalent to $\sim p \leftrightarrow \sim q$?

Lesson 2.2

LESSON 2.3

TEACHER'S NAME ______ CLASS ______ ROOM ______ DATE ______

Lesson Plan

2-day lesson (See *Pacing the Chapter,* TE pages 68C–68D)

For use with pages 86–95

GOALS

1. Use symbolic notation to represent logical statements.
2. Form conclusions by applying the laws of logic to true statements.

State/Local Objectives ______

______

✓ Check the items you wish to use for this lesson.

STARTING OPTIONS

____ Homework Check: TE page 82: Answer Transparencies
____ Warm-Up or Daily Homework Quiz: TE pages 87 and 85, CRB page 40, or Transparencies

TEACHING OPTIONS

____ Motivating the Lesson: TE page 88
____ Concept Activity: SE page 86; CRB page 41 (Activity Support Master)
____ Lesson Opener (Application): CRB page 42 or Transparencies
____ Examples: Day 1: 1–6, SE pages 87–90; Day 2: See the Extra Examples.
____ Extra Examples: Day 1 or Day 2: 1–6, TE pages 88–90 or Transp.; Internet
____ Closure Question: TE page 90
____ Guided Practice: SE page 91 Day 1: Exs. 1–7; Day 2: See Checkpoint Exs. TE pages 88–90

APPLY/HOMEWORK

Homework Assignment

____ Basic Day 1: 8–42 even, 45–48; Day 2: 9–41 odd, 51, 56–62 even; Quiz 1–5
____ Average Day 1: 8–42 even, 45–48; Day 2: 9–41 odd, 49–51, 56–62 even; Quiz 1: 1–5
____ Advanced Day 1: 8–42 even, 45–48; Day 2: 9–41 odd, 44, 49–55, 56–62 even; Quiz 1: 1–5

Reteaching the Lesson

____ Practice Masters: CRB pages 43–45 (Level A, Level B, Level C)
____ Reteaching with Practice: CRB pages 46–47 or Practice Workbook with Examples
____ Personal Student Tutor

Extending the Lesson

____ Applications (Interdisciplinary): CRB page 49
____ Math and History: SE page 95; CRB page 50; Internet
____ Challenge: SE page 94; CRB page 51 or Internet

ASSESSMENT OPTIONS

____ Checkpoint Exercises: Day 1 or Day 2: TE page 88–90 or Transparencies
____ Daily Homework Quiz (2.3): TE page 94, CRB page 55, or Transparencies
____ Standardized Test Practice: SE page 94; TE page 94; STP Workbook; Transparencies
____ Quiz (2.1–2.3): SE page 95; CRB page 52

Notes ______

______

______

LESSON 2.3

TEACHER'S NAME ______________________ CLASS ________ ROOM ________ DATE ________

Lesson Plan for Block Scheduling

1-day lesson (See *Pacing the Chapter,* TE pages 68C–68D)   For use with pages 86–95

GOALS

1. **Use symbolic notation to represent logical statements.**
2. **Form conclusions by applying the laws of logic to true statements.**

State/Local Objectives ____

____

____

CHAPTER PACING GUIDE	
Day	**Lesson**
1	Assess Ch. 1; 2.1 (all)
2	2.2 (all)
3	**2.3 (all)**
4	2.4 (all); 2.5 (begin)
5	2.5 (end); 2.6 (begin)
6	2.6 (end); Review Ch. 2
7	Assess Ch. 2; 3.1 (all)

✓ Check the items you wish to use for this lesson.

STARTING OPTIONS

____ Homework Check: TE page 82: Answer Transparencies
____ Warm-Up or Daily Homework Quiz: TE pages 87 and 85, CRB page 40, or Transparencies

TEACHING OPTIONS

____ Motivating the Lesson: TE page 88
____ Concept Activity: SE page 86; CRB page 41 (Activity Support Master)
____ Lesson Opener (Application): CRB page 42 or Transparencies
____ Examples: 1–6, SE pages 87–90
____ Extra Examples: TE pages 88–90 or Transparencies; Internet
____ Closure Question: TE page 90
____ Guided Practice Exercises: SE page 91

APPLY/HOMEWORK

Homework Assignment

____ Block Schedule: 8–42, 45–51; 56–62 even; Quiz 1: 1–5

Reteaching the Lesson

____ Practice Masters: CRB pages 43–45 (Level A, Level B, Level C)
____ Reteaching with Practice: CRB pages 46–47 or Practice Workbook with Examples
____ Personal Student Tutor

Extending the Lesson

____ Applications (Interdisciplinary): CRB page 49
____ Math and History: SE page 95; CRB page 50; Internet
____ Challenge: SE page 94; CRB page 51 or Internet

ASSESSMENT OPTIONS

____ Checkpoint Exercises: TE pages 88–90 or Transparencies
____ Daily Homework Quiz (2.3): TE page 94, CRB page 55, or Transparencies
____ Standardized Test Practice: SE page 94; TE page 94; STP Workbook; Transparencies
____ Quiz (2.1–2.3): SE page 95; CRB page 52

Notes ____

____

____

LESSON 2.3

NAME ______________________ DATE ________

Available as a transparency

Warm-Up Exercises

For use before Lesson 2.3, pages 86–95

Write the contrapositive of each statement.

1. If it thunders, the golf tournament is canceled.

2. If $x(x - 1) = 0$, then $x = 0$ or 1.

3. If $m\angle A = 45°$, then $\angle A$ is congruent to $\angle B$.

Daily Homework Quiz

For use after Lesson 2.2, pages 79–85

Rewrite the biconditional statement as a conditional statement and its converse.

1. We will go to the beach if and only if it is sunny.

Give a counterexample that demonstrates that the converse of the statement is false.

2. If a polygon has four equal sides, then it is a square.

3. If a vehicle has wheels, then it is a car.

Determine whether the statement can be combined with its converse to form a true biconditional.

4. If $2x > 8$, then $x = 5$.

NAME ____ DATE ______________

Activity Support Master

For use with page 86

		Hometown					Hobbies				
		Hart's Location	Grand Rapids	Stockton	Ravenna	Springdale	Basketball	Reading	Playing computer games	Playing the guitar	In-line skating
Students	Maynard										
	Tamara										
	Dave										
	Marie										
	Brad										
Hobbies	Basketball										
	Reading										
	Playing computer games										
	Playing the guitar										
	In-line skating										

NAME _______________________________ DATE ____________

Available as a transparency

Activity Lesson Opener

For use with pages 87–95

SET UP: Work with a partner.

YOU WILL NEED: • index cards

Consider the following statements.

Statement A: The student is a boy.

Statement B: The student is a girl.

1. Write Statement A on one index card, Statement B on another, the word "not" on each of two index cards, and an arrow, $\rightarrow$, on another index card. The arrow, $\rightarrow$, means "implies."

2. Arrange the cards to make as many conditional statements as you can. After you arrange a statement, record it using symbols. For example, the following arrangement is recorded as $A \rightarrow B$.

Statement A: The student is a boy.	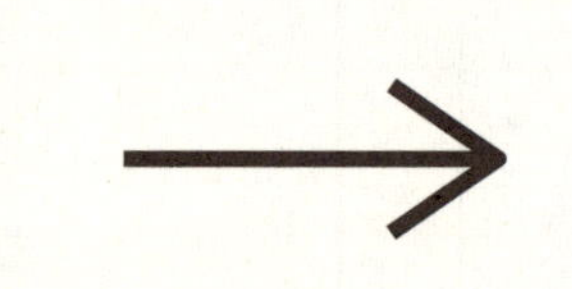	Statement B: The student is a girl.

Then decide whether the statement is *true* or *false*.

3. Did you create any statements that could be combined to form a biconditional statement? If so, give the pair(s) of statements that compose the biconditional(s).

4. How do you think a double arrow, $\leftrightarrow$, could be used to write a biconditional using symbolic notation? Use your answer to write the biconditional statement(s) using symbols. Decide whether each statement you created is *true* or *false* and explain your answer.

LESSON 2.3

NAME ______________________ DATE ________

Practice A

For use with pages 87–95

Using p and q below, write the symbolic statement in words.

p: The sky is cloudy. q: It is raining.

1. $\sim p$
2. $\sim q$
3. $q \rightarrow p$
4. $\sim q \rightarrow \sim p$
5. $p \rightarrow q$
6. $\sim p \rightarrow \sim q$

Determine if statement (3) follows from statements (1) and (2) by the Law of Detachment or the Law of Syllogism. If it does, state which law was used. If it does not, write invalid.

7. (1) If an angle is acute, then it is not obtuse.
   (2) $\angle ABC$ is acute.
   (3) $\angle ABC$ is not obtuse.
8. (1) Right angles are congruent.
   (2) $\angle A \cong \angle B$
   (3) $\angle A$ and $\angle B$ are right angles.
9. (1) If you save a penny, then you have earned a penny.
   (2) Art saves a penny.
   (3) Art has earned a penny.
10. (1) If you are a teenager, then you are always right.
    (2) If you are always right, then people will listen to you.
    (3) If you are a teenager, then people will listen to you.
11. (1) If you drive 50 miles per hour in a school zone, then you will get a speeding ticket.
    (2) Pat received a speeding ticket.
    (3) Pat was driving 50 miles per hour in a school zone.
12. (1) If $m\angle 2 = 40°$, then $m\angle 3 = 140°$.
    (2) If $m\angle 3 = 140°$, then $m\angle 4 = 40°$.
    (3) If $m\angle 2 = 40°$, then $m\angle 4 = 40°$.

Write a conclusion using the true statements. If no conclusion is possible, write no conclusion.

13. If Tim gets stung by a bee, then he will get very ill. If he gets very ill, then he will go to the hospital. Tim gets stung by a bee.
14. If Hank applies for the job, then he will be the new lifeguard at the pool. If he is the new lifeguard at the pool, then he will buy a new car. Hank applies for the job.
15. If two planes intersect, then their intersection is a line. Plane A and plane B intersect.
16. If you cut class, then you will receive a detention. You cut class.
17. If Jay doesn't work hard, then he won't start the game and will quit the team. Jay quit the team.
18. If B is between A and C, then $AB + BC = AC$. $AB = 3$ cm, $BC = 2$ cm.

Lesson 2.3

NAME ______________________________ DATE __________

Practice B

For use with pages 87–95

Using *p* and *q* below, write the symbolic statement in words.

p: The value of x is 4. q: $3x + 2 = 14$

1. $\sim p$ **2.** $\sim q$ **3.** $q \rightarrow p$

4. $\sim q \rightarrow \sim p$ **5.** $p \rightarrow q$ **6.** $\sim p \rightarrow \sim q$

Determine if statement (3) follows from statements (1) and (2) by the Law of Detachment or the Law of Syllogism. If it does, state which law was used. If it does not, write invalid.

7. (1) If an angle measures more than 90°, then it is not acute.
(2) $m\angle ABC = 120°$
(3) $\angle ABC$ is not acute.

8. (1) All 45° angles are congruent.
(2) $\angle A \cong \angle B$
(3) $\angle A$ and $\angle B$ are 45° angles.

9. (1) If you order the apple pie, then it will be served with ice cream.
(2) Matthew ordered the apple pie.
(3) Matthew was served ice cream.

10. (1) If you wear the school colors, then you have school spirit.
(2) If you have school spirit, then the team feels great.
(3) If you wear the school colors, then the team will feel great.

11. (1) If you eat too much turkey, then you will get sick.
(2) Kinsley got sick.
(3) Kinsley ate too much turkey.

12. (1) If $\angle 2$ is acute, then $\angle 3$ is obtuse.
(2) If $\angle 3$ is obtuse, then $\angle 4$ is acute.
(3) If $\angle 2$ is acute, then $\angle 4$ is acute.

In Exercises 13–18, assume the following statements are true.

- If Susan screams, then the dog will run away.
- If the dog licks Susan, then Susan will scream.
- If Carl tells the dog to "kiss," then the dog will lick Susan.
- Carl tells the dog to "kiss."

13. Write the contrapositive of the third statement.

14. Write the inverse of the second statement.

15. Write the converse of the first statement.

16. Write the premises in an order which makes a valid argument.

17. Did the dog lick Susan? Explain.

18. Did the dog run away? Explain.

LESSON 2.3

NAME ______________________________ DATE __________

Practice C

For use with pages 87–95

Using *p* and *q* below, write the symbolic statement in words. Assume *p* and *q* are true. Decide if each symbolic statement is *true* or *false*.

p: The value of x is -4. q: $3x + 2 = -10$

1. $\sim p$ **2.** $\sim q$ **3.** $q \rightarrow p$

4. $\sim q \rightarrow \sim p$ **5.** $p \rightarrow q$ **6.** $\sim p \rightarrow \sim q$

Determine if statement (3) follows from statements (1) and (2) by the Law of Detachment or the Law of Syllogism. If it does, state which law was used. If it does not, write invalid.

7. (1) If you are a student, then you have lots of homework.
(2) If you have lots of homework, then you have no social life.
(3) If you are a student, then you have no social life.

8. (1) If the lines are perpendicular, then they intersect to form a right angle.
(2) Line l is perpendicular to line m.
(3) Lines l and m intersect to form a right angle.

9. (1) Vertical angles are congruent.
(2) $\angle A \cong \angle B$
(3) $\angle A$ and $\angle B$ are vertical angles.

10. (1) If the quadrilateral is a square, then it has four right angles.
(2) Quadrilateral *ABCD* has four right angles.
(3) Quadrilateral *ABCD* is a square.

11. (1) If you practice your clarinet, then you will improve.
(2) Kevin practices his clarinet.
(3) Kevin's clarinet playing improves.

12. (1) If $m\angle 2 \neq 40°$, then $m\angle 3 \neq 140°$.
(2) If $m\angle 3 \neq 140°$, the $m\angle 4 \neq 40°$.
(3) If $m\angle 2 \neq 40°$, then $m\angle 4 \neq 40°$.

In Exercises 13–17, assume the following statements are true.

- If I call the supervisor, then I must pay the rent. $(p \rightarrow q)$
- If the apartment ceiling is not leaking, then it is not raining. $(\sim r \rightarrow \sim s)$
- I will call the supervisor if the apartment ceiling leaks. $(p \rightarrow r)$
- If it is not raining, then it is not Tuesday. $(\sim s \rightarrow \sim t)$

13. Write the contrapositive of the second statement.

14. Write the contrapositive of the fourth statement.

15. Write the premises in an order which makes a valid argument.

16. It is Tuesday. Can you conclude that you must pay the rent? Explain.

17. It is not Tuesday. Can you conclude you did not pay the rent? Explain.

LESSON 2.3

NAME ______________________ DATE ________

Reteaching with Practice

For use with pages 87–95

GOAL **Use symbolic notation to represent logical statements and form conclusions by applying the laws of logic to true statements**

VOCABULARY

Law of Detachment

If $p \rightarrow q$ is a true conditional statement and p is true, then q is true.

Law of Syllogism

If $p \rightarrow q$ and $q \rightarrow r$ are true conditional statements, then $p \rightarrow r$ is true.

EXAMPLE 1 *Using Symbolic Notation*

Let p be "the value of x is 3" and let q be "$x^3 = 27$."

a. Write $p \rightarrow q$ in words. (Note: $\rightarrow$ is read as "implies.")

b. Write $q \rightarrow p$ in words.

c. Decide whether the biconditional statement $p \leftrightarrow q$ is true. (*Note:* $\leftrightarrow$ is read as "if and only if.")

SOLUTION

a. If the value of x is 3, then $x^3 = 27$.

b. If $x^3 = 27$, then the value of x is 3.

c. The conditional statement in (a) is true and its converse in (b) is also true. So, the biconditional statement $p \leftrightarrow q$ is true.

Exercises for Example 1

Use p and q to (a) write $p \rightarrow q$ in words, (b) write $q \rightarrow p$ in words, and (c) decide whether the biconditional statement $p \leftrightarrow q$ is true.

1. p: two lines are parallel
q: two lines do not intersect

2. p: you are in North America
q: you are in the United States

EXAMPLE 2 *Writing an Inverse and a Contrapositive*

Let p be "a word is misspelled" and q be "a word is not in the dictionary."

a. Write the contrapositive of $p \rightarrow q$.

b. Write the inverse of $p \rightarrow q$.

SOLUTION

a. Contrapositive: $\sim q \rightarrow \sim p$
If a word is in the dictionary, then it is not misspelled.

b. Inverse: $\sim p \rightarrow \sim q$
If a word is not misspelled, then it is in the dictionary.

Lesson 2.3

NAME ______________________________ DATE __________

Reteaching with Practice

For use with pages 87–95

Exercises for Example 2

Use p and q to (a) write the contrapositive of $p \rightarrow q$ and (b) write the inverse of $p \rightarrow q$.

3. p: the measures of two angles sum to 180°

q: two angles are supplementary

4. p: a number is divisible by 10

q: a number is divisible by 5

EXAMPLE 3 Using the Laws of Deductive Reasoning

Use the true statements to determine whether the conclusion is *true* or *false.*

- If it looks like rain, then I will bring my umbrella to school with me.
- If there are clouds in the sky and the sky is dark, then it looks like rain.
- If I bring my umbrella to school with me, then I will hang it in the classroom closet.
- This morning, there are clouds in the sky and the sky is dark.

Conclusion: My umbrella is hanging in the classroom closet.

Solution

The conclusion is *true.* Let p, q, r, and s represent the following:

p: There are clouds in the sky and the sky is dark.

q: It looks like rain.

r: I bring my umbrella to school with me.

s: My umbrella is hanging in the classroom closet.

p is true, as provided by the fourth point. Because $p \rightarrow q$ is true (second point), $q \rightarrow r$ is true (first point), and $r \rightarrow s$ is true (third point), then $p \rightarrow s$ is true by the Law of Syllogism.

You are given that there are clouds in the sky and that the sky is dark, so p is true. Using the Law of Detachment, you can conclude that my umbrella is hanging in the classroom closet.

Exercises for Example 3

5. Use the true statements to determine whether the conclusion is true or false. Explain your reasoning.

- If Pete's thirsty, he will want to drink a milkshake.
- If Pete wants to drink a milkshake, he will need to use the blender.
- If Pete needs the blender, he will need to clean it first.
- The blender is dirty.

Conclusion: Pete needs to clean the blender.

LESSON 2.3

NAME ______________________________ DATE ________

Quick Catch-Up for Absent Students

For use with pages 86–95

The items checked below were covered in class on (date missed) __________

Activity 2.3: Logic Puzzle (p. 86)

____ **Goal:** Use deductive reasoning to solve a logic puzzle.

Lesson 2.3: Deductive Reasoning

____ **Goal 1:** Use symbolic notation to represent logical statements. (pp. 87–88)

Material Covered:

____ Example 1: Using Symbolic Notation

____ Student Help: Study Tip

____ Example 2: Writing an Inverse and a Contrapositive

____ **Goal 2:** Form conclusions by applying the laws of logic to true statements. (pp. 89–90)

Material Covered:

____ Example 3: Using Inductive and Deductive Reasoning

____ Example 4: Using the Law of Detachment

____ Example 5: Using the Law of Syllogism

____ Example 6: Using the Laws of Deductive Reasoning

Vocabulary:

logical argument, p. 89

Law of Detachment, p. 89

Law of Syllogism, p. 89

____ Other (specify) ____

____

Homework and Additional Learning Support

____ Textbook (specify) pp. 91–95

____

____ Internet: Extra Examples at www.mcdougallittell.com

____ *Reteaching with Practice* worksheet (specify exercises) __________

____ *Personal Student Tutor* for Lesson 2.3

Lesson 2.3

NAME ______________________________ DATE ____________

Interdisciplinary Application

For use with pages 87–95

Baseball

PHYSICAL EDUCATION You are learning the basic rules of baseball in gym class. Your gym teacher explains that teams consist of nine people who use a hard ball, a wooden or aluminum bat, and gloves. Teams alternate turns in the field and at bat. One turn at each constitutes an inning, and nine innings constitute a game. In the field there is a pitcher, a catcher, four infielders, and three outfielders. The pitcher throws overhand from the pitching rubber to home plate, while an opposing batter attempts to hit the pitches and safely reach base. If the batter hits the ball, the fielders attempt to put the batter out through various plays.

A run is scored every time a batter crosses home plate after touching each base in the prescribed order. When the fielding team puts out three batters, the two teams exchange places. A team wins when it scores more than the opposing team at the end of nine innings. Extra innings are played if the score is tied.

In Exercises 1–3, use the following conditional statement.

If the batter makes an out, then the batter does not become a base runner.

1. Write the converse of the conditional statement.

2. Write the contrapositive of the conditional statement.

3. Write the inverse of the conditional statement.

In Exercises 4–6, use the following contrapositive.

If play does not continue into extra innings, then the score is not tied at the end of nine innings.

4. Write the conditional statement.

5. Write the converse.

6. Write the inverse.

7. If your team scores a run, then your team will win the game. Using the Law of Detachment, what can you conclude when your team scores a run?

8. Write some conditional statements that can be made from the following true statements using the Law of Syllogism.

a. If a batter crosses home plate, then a run is scored.

b. If there are three outs, then the inning is over.

c. If the outfielder makes the catch, then there will be three outs.

d. If a run is scored, then the game will be tied.

NAME ____________________ DATE __________

Math and History Application

For use with page 95

HISTORY Charles Dodgson (1832–1898) is better known by his pseudonym—Lewis Carroll. Carroll was the third of eleven children, all of whom stuttered. He was uneasy in the presence of adults, and could speak without stuttering only to small children. Carroll could easily relate to young children. Not surprisingly, he wrote a number of children's books, including *Alice in Wonderland* and *Through the Looking Glass.* Carroll was also ordained in the Church of England in 1861, but never practiced his ministry.

MATH To mathematicians, however, Carroll is best known for his accomplishments in the field of logic. In addition to articles and books on geometry, Carroll is also the author of numerous mathematical puzzles, including the following:

We are offered our choice of two clocks.
Clock A loses 1 minute per day.
Clock B doesn't run at all.

A

B

Which clock should we choose?

1. Once Clock A is properly set, how many minutes must it lose before it shows the correct time again?
2. How many days will it take Clock A to lose the number of minutes in your answer to Exercise 1?
3. If Clock A is properly set today, when will it show the correct time again (in years)?
4. How often is Clock B correct? Which clock should we choose?

NAME ______________________ DATE __________

Challenge: Skills and Applications

For use with pages 87–95

1. In a mathematics class, Atsuko is seated in front of Bradley, and Bradley is seated in front of Consuela. The teacher said, "I have 3 yellow hats and 2 blue hats. I have placed 3 of these 5 hats on your heads. Can you determine what color hat you have on your own head?
   a. Consuela correctly states that she cannot determine the color of her hat. What conclusion can the other students draw?
   b. Bradley, who has heard Consuela's answer but cannot see Consuela, correctly states that he cannot deduce the color of his hat. Then Atsuko, who cannot see the other two students, says, "Now I know what color my hat is!" What color is Atsuko's hat? How does she know?
   c. Can you deduce the color of Bradley's hat or Consuela's hat?
2. Three married couples are on vacation together. Use the following information to determine the full name and home of each individual.

   The men's names are Ivan, Jeff, and Kevin.

   The women's names are Linda, Mary, and Nancy.

   Their last names are Ricardo, Smith, and Tanaka.

   Their homes are Boston, Chicago, and Denver.

   Nancy does not live in Boston.

   Neither Kevin nor Mary lives in Denver.

   Ivan Tanaka lives in Boston.

   Linda Ricardo does not live in Denver.
3. A student is playing a game that begins with three coins, each with heads showing. At each turn, the student turns over exactly two of the three coins.
   a. After the first turn, how many heads are showing?
   b. Suppose an odd number of heads are showing. Explain why taking one turn cannot produce an even number of heads.
   c. Using the result from part (b), explain why the student will *never* get exactly 2 heads showing.

LESSON 2.3

NAME ______________________ DATE __________

Quiz 1

For use after Lessons 2.1–2.3

Write the true statement in if-then form and write its converse. Determine whether the statement and its converse can be combined to form a true biconditional statement. *(Lesson 2.1 and Lesson 2.2)*

1. If this month is September, then next month is October.
2. A decade is a period of 10 years.
3. I breathe when I sleep.

Logical Reasoning **Use the true statements to answer the questions.** *(Lesson 2.3)*

- If you win the football game, then the coach will be happy.
- If your team scores 4 touchdowns, then it will win the game.
- Your team scores 4 touchdowns.

4. Does your team win?
5. Is the coach happy?

Answers

1. ______________________

2. ______________________

3. ______________________

4. ______________________

5. ______________________

Lesson 2.3

LESSON

TEACHER'S NAME ______________________ CLASS ________ ROOM ______ DATE ______

Lesson Plan

1-day lesson (See *Pacing the Chapter,* TE pages 68C–68D) **For use with pages 96–101**

GOALS

1. **Use properties from algebra.**
2. **Use properties of length and measure to justify segment and angle relationships.**

State/Local Objectives ____

____

✓ Check the items you wish to use for this lesson.

STARTING OPTIONS

____ Homework Check: TE page 91: Answer Transparencies
____ Warm-Up or Daily Homework Quiz: TE pages 96 and 94, CRB page 55, or Transparencies

TEACHING OPTIONS

____ Lesson Opener (Activity): CRB page 56 or Transparencies
____ Examples: 1–5, SE pages 96–98
____ Extra Examples: TE pages 97–98 or Transparencies; Internet
____ Closure Question: TE page 98
____ Guided Practice Exercises: SE page 99

APPLY/HOMEWORK

Homework Assignment

____ Basic 10–15, 16–26 even, 32, 36–50 even
____ Average 10–15, 16–28 even, 32, 36–50 even
____ Advanced 10–15, 16–26 even, 28–34, 36–50 even

Reteaching the Lesson

____ Practice Masters: CRB pages 57–59 (Level A, Level B, Level C)
____ Reteaching with Practice: CRB pages 60–61 or Practice Workbook with Examples
____ Personal Student Tutor

Extending the Lesson

____ Cooperative Learning Activity: CRB page 63
____ Applications (Real-Life): CRB page 64
____ Challenge: SE page 101; CRB page 65 or Internet

ASSESSMENT OPTIONS

____ Checkpoint Exercises: TE pages 97–98 or Transparencies
____ Daily Homework Quiz (2.4): TE page 101, CRB page 68, or Transparencies
____ Standardized Test Practice: SE page 101; TE page 101; STP Workbook; Transparencies

Notes ____

____

____

LESSON 2.4

TEACHER'S NAME ______________________ CLASS ________ ROOM ______ DATE ______

Lesson Plan for Block Scheduling

Half-day lesson (See *Pacing the Chapter,* TE pages 68C–68D) For use with pages 96–101

GOALS

1. **Use properties from algebra.**
2. **Use properties of length and measure to justify segment and angle relationships**

State/Local Objectives ______________________________

____

____

CHAPTER PACING GUIDE	
Day	**Lesson**
1	Assess Ch. 1; 2.1 (all)
2	2.2 (all)
3	2.3 (all)
4	**2.4 (all)**; 2.5 (begin)
5	2.5 (end); 2.6 (begin)
6	2.6 (end); Review Ch. 2
7	Assess Ch. 2; 3.1 (all)

✓ Check the items you wish to use for this lesson.

STARTING OPTIONS

____ Homework Check: TE page 91: Answer Transparencies

____ Warm-Up or Daily Homework Quiz: TE pages 96 and 94, CRB page 55, or Transparencies

TEACHING OPTIONS

____ Lesson Opener (Activity): CRB page 56 or Transparencies

____ Examples: 1–5, SE pages 96–98

____ Extra Examples: TE pages 97–98 or Transparencies; Internet

____ Closure Question: TE page 98

____ Guided Practice: SE page 99

APPLY/HOMEWORK

Homework Assignment (See also the assignment for Lesson 2.5.)

____ Block Schedule: 10–15, 16–28 even, 32, 36–50 even

Reteaching the Lesson

____ Practice Masters: CRB pages 57–59 (Level A, Level B, Level C)

____ Reteaching with Practice: CRB pages 60–61 or Practice Workbook with Examples

____ Personal Student Tutor

Extending the Lesson

____ Cooperative Learning Activity: CRB page 63

____ Applications (Real-Life): CRB page 64

____ Challenge: SE page 101; CRB page 65 or Internet

ASSESSMENT OPTIONS

____ Checkpoint Exercises: TE pages 97–98 or Transparencies

____ Daily Homework Quiz (2.4): TE page 101, CRB page 68, or Transparencies

____ Standardized Test Practice: SE page 101; TE page 101; STP Workbook; Transparencies

Notes ____

____

____

LESSON **2.4**

NAME ______________________ DATE __________

Available as a transparency

Warm-Up Exercises

For use before Lesson 2.4, pages 96–101

Solve each equation.

1. $3x = 27$

2. $x + 6 = -17$

3. $x - 9 = 18$

4. $\frac{2}{3}x = 6$

5. $-x = 4$

Daily Homework Quiz

For use after Lesson 2.3, pages 86–95

Using *p* and *q* below, write each statement as directed.

p: This animal has stripes.

q: This animal is a tiger.

1. $\sim p$

2. Write the inverse of $p \rightarrow q$ in symbols and words.

3. Write the contrapositive of $p \rightarrow q$ in symbols and words.

Use the Law of Syllogism to write the statement that follows from the pair of true statements.

4. If it is Friday, then tomorrow is Saturday. If it is Saturday, we will swim.

Lesson 2.4

LESSON 2.4

Available as a transparency

NAME ______________________ DATE __________

Activity Lesson Opener

For use with pages 96–101

SET UP: Work with a partner.

1. The steps to solve the equation $10y + 5 = 25$ are given below. Put the steps in the correct order. Next to each step, describe what happens in that step.

$$10y + 5 - 5 = 25 - 5$$

$$y = 2$$

$$10y = 20$$

$$\frac{10y}{10} = \frac{20}{10}$$

$$10y + 5 = 25$$

2. The steps to solve the equation $6x + 3 = 9(x - 1)$ are given below. Put the steps in the correct order. Next to each step, describe what happens in that step.

$$-3 + 3 = -9$$

$$6x + 3 = 9x - 9$$

$$-3x = -12$$

$$6x + 3 = 9(x - 1)$$

$$\frac{-3x}{-3} = \frac{-12}{-3}$$

$$6x - 9x + 3 = 9x - 9x - 9$$

$$x = 4$$

$$-3x + 3 - 3 = -9 - 3$$

Lesson 2.4

LESSON **2.4**

NAME ______________________________ DATE __________

Practice A

For use with pages 96–101

Match the statement with the Property of Equality.

1. If $JK = PQ$ and $PQ = ST$, then $JK = ST$.
2. If $m\angle S = 30°$, then $5° + m\angle S = 35°$.
3. If $AB + CD = EF + CD$, then $AB = EF$.
4. $AB = AB$
5. If $ST = 2$, then $ST + TU = 2 + TU$.
6. If $m\angle K = 45°$, then $3(m\angle K) = 135°$.
7. If $m\angle P = m\angle Q$, then $m\angle Q = m\angle P$.

A. Addition property
B. Reflexive property
C. Substitution property
D. Transitive property
E. Symmetric property
F. Multiplication property
G. Subtraction property

In Exercises 8–13, use the property to complete the statement.

8. Addition property of equality: If $AB = 5$, then $10 + AB =$ __?__.
9. Multiplication property of equality: If $m\angle C = 30°$, then __?__ $(m\angle C) = 15°$.
10. Reflexive property of equality: $AF =$ __?__.
11. Symmetric property of equality: If $m\angle DCF = m\angle MJC$, then __?__.
12. Transitive property of equality: If $YZ = DB$ and __?__ $= JK$, then __?__.
13. Substitution property of equality: If $MN = 3$, then $5(MN) =$ __?__.

Complete the argument, giving a reason for each step.

14. $3(2x - 4) = 5x + 2$ Given
 $6x - 12 = 5x + 2$ **a.** __?__
 $x - 12 = 2$ **b.** __?__
 $x = 14$ **c.** __?__

15. $4x + 8 = 2x - 12$ Given
 $2x + 8 = -12$ **a.** __?__
 $2x = -20$ **b.** __?__
 $x = -10$ **c.** __?__

16. $m\angle AEB + m\angle BEC = m\angle CED + m\angle BEC$ Given
 $m\angle BEC = m\angle BEC$ **a.** __?__
 $m\angle AEB = m\angle CED$ **b.** __?__

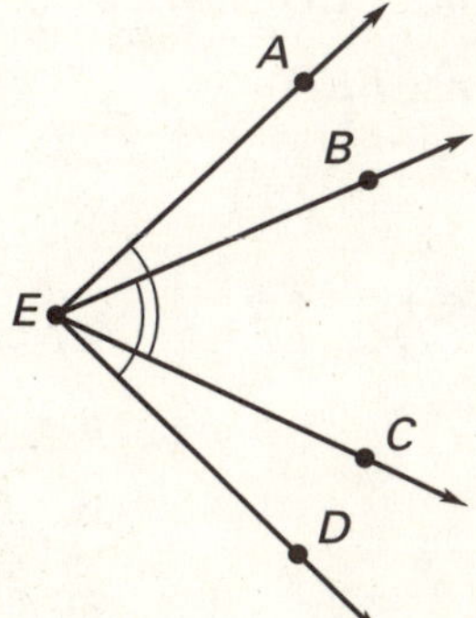

In Exercises 17 and 18, solve the equation and state a reason for each step.

17. $5(2x - 1) = 9x + 4$
18. $-4(x - 5) = 13$

Lesson 2.4

LESSON

2.4

NAME ______________________ DATE ________

Practice B

For use with pages 96–101

In Exercises 1–6, use the property to complete the statement.

1. Reflexive property of angle measure: $m\angle T =$ __?__.
2. Transitive property of equality: If $KL = MN$ and __?__ $= RW$, then __?__.
3. Addition property of equality: If $x = 5$, then $17 + x =$ __?__.
4. Symmetric property of equality: If $BC = RL$, then __?__.
5. Substitution property of equality: If $m\angle A = 45°$, then $3(m\angle A) =$ __?__.
6. Multiplication property of equality: If $m\angle A = 45°$, then __?__ $(m\angle A) = 15°$.

Complete the argument, giving a reason for each step.

7.

$5(2x - 1) = 9x + 2$	Given
$10x - 5 = 9x + 2$	**a.** __?__
$10x = 9x + 7$	**b.** __?__
$x = 7$	**c.** __?__

8.

$8x - 5 = -2x - 15$	Given
$10x - 5 = -15$	**a.** __?__
$10x = -10$	**b.** __?__
$x = -1$	**c.** __?__

9.

$AB = BC$	Given
$AC = AB + BC$	**a.** __?__
$AC = AB + AB$	**b.** __?__
$AC = 2(AB)$	**c.** __?__

A B C

10.

$m\angle AEB = m\angle CED$	Given
$m\angle BEC = m\angle BEC$	**a.** __?__
$m\angle AEB + m\angle BEC = m\angle CED + m\angle BEC$	**b.** __?__
$m\angle AEC = m\angle AEB + m\angle BEC$	**c.** __?__
$m\angle BED = m\angle CED + m\angle BEC$	**d.** __?__
$m\angle AEC = m\angle BED$	**e.** __?__

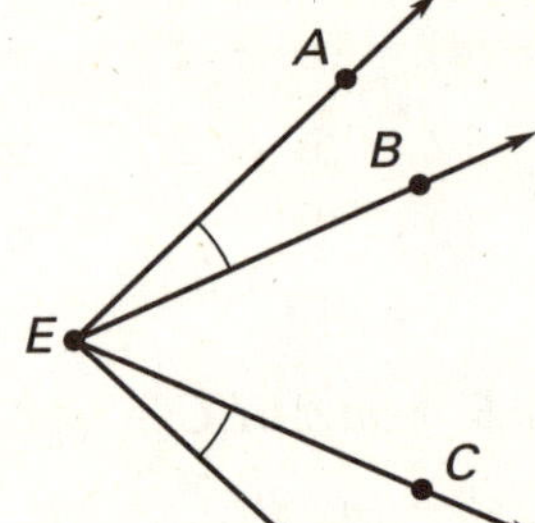

11.

$\overleftrightarrow{AB} \perp \overleftrightarrow{EF}$, $\overleftrightarrow{CD} \perp \overleftrightarrow{EF}$	Given
$m\angle 1 = 90°$	**a.** __?__
$m\angle 2 = 90°$	**b.** __?__
$m\angle 1 = m\angle 2$	**c.** __?__

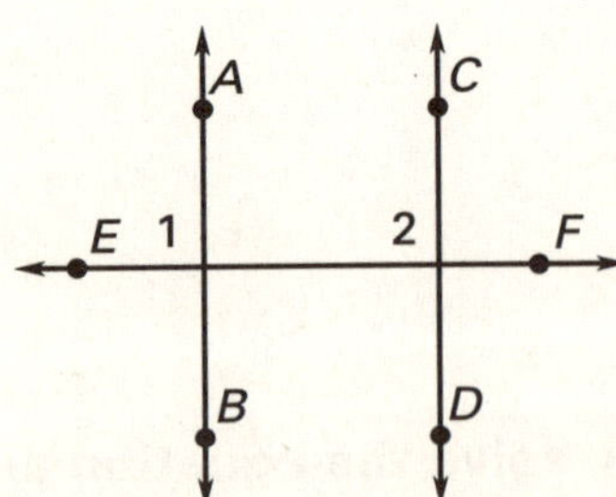

LESSON 2.4

NAME ______________________________ DATE __________

Practice C

For use with pages 96–101

In Exercises 1–6, use the property to complete the statement.

1. Multiplication property of equality: If $BD = 6$, then ___?___ $(BD) = 2$.
2. Transitive property of equality: If $m\angle ABC = m\angle DEF$ and ___?___ $= m\angle STO$, then ___?___.
3. Addition property of equality: If $m\angle STJ = 42°$, then $17° + m\angle STJ =$ ___?___.
4. Reflexive property of equality: $HS =$ ___?___.
5. Substitution property of equality: If $RL = 7.5$, then $5(RL) =$ ___?___.
6. Symmetric property of equality: If $m\angle MCS = m\angle DBA$, then ___?___.

Complete the argument, giving a reason for each step.

7.

Statement	Reason
$-2(3x - 4) = 3x + 12$	Given
$-6x + 8 = 3x + 12$	a. ___?___
$-9x + 8 = 12$	b. ___?___
$-9x = 4$	c. ___?___
$x = -\frac{4}{9}$	d. ___?___

8.

Statement	Reason
$9 = 4x - 3(x - 2)$	Given
$9 = 4x - 3x + 6$	a. ___?___
$9 = x + 6$	b. ___?___
$3 = x$	c. ___?___
$x = 3$	d. ___?___

9.

Statement	Reason
$AB = CD$	Given
$BC = BC$	a. ___?___
$AB + BC = CD + BC$	b. ___?___
$AB + BC = AC$	c. ___?___
$CD + BC = BD$	d. ___?___
$AC = BD$	e. ___?___

A B C D

In Exercises 10 and 11, give an argument for the statement, including a reason for each step.

10. If $\overleftrightarrow{AB} \perp \overleftrightarrow{EF}$, $\overleftrightarrow{CD} \perp \overleftrightarrow{EF}$, then $m\angle 1 = m\angle 2$.

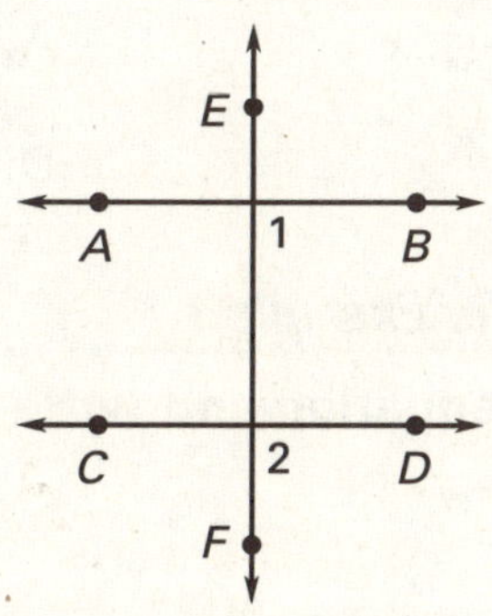

11. If $m\angle 1 = 40°$ and $m\angle 2 = 50°$, then the angles are complementary.

Lesson 2.4

LESSON 2.4

NAME ______________________ DATE __________

Reteaching with Practice

For use with pages 96–101

GOAL Use properties from algebra and use properties of length and measure to justify segment and angle relationships

Vocabulary

Algebraic Properties of Equality
Let a, b, and c be real numbers.

Addition Property If $a = b$, then $a + c = b + c$.

Subtraction Property If $a = b$, then $a - c = b - c$.

Multiplication Property If $a = b$, then $ac = bc$.

Division Property If $a = b$ and $c \neq 0$, then $a \div c = b \div c$.

Reflexive Property For any real number a, $a = a$.

Symmetric Property If $a = b$, then $b = a$.

Transitive Property If $a = b$ and $b = c$, then $a = c$.

Substitution Property If $a = b$, then a can be substituted for b in any equation or expression.

EXAMPLE 1 Writing Reasons

Solve $10 - 2x = 3(x - 2) + 4$ and write a reason for each step.

Solution

$10 - 2x = 3(x - 2) + 4$	Given
$10 - 2x = 3x - 6 + 4$	Distributive property
$10 - 2x = 3x - 2$	Simplify.
$12 - 2x = 3x$	Addition property of equality
$12 = 5x$	Addition property of equality
$\frac{12}{5} = x$	Division property of equality

Exercises for Example 1

Solve the equation and write a reason for each step.

1. $2x + 3 = 7x$

2. $4 + 2(3x + 5) = 11 - x$

3. $6x - 2 = -4(x - 1)$

4. $\frac{1}{5}x + 4 = 2x + \frac{3}{5}$

Lesson 2.4

EXAMPLE 2 Using Properties of Length and Measure

In the diagram, $WY = XZ$, show that $WX = YZ$.

W X Y Z

NAME ______________________________ DATE ____________

Reteaching with Practice

For use with pages 96–101

Solution

$WY = XZ$	Given
$WY = WX + XY$	Segment Addition Postulate
$XZ = XY + XZ$	Segment Addition Postulate
$WX + XY = XY + YZ$	Substitution property of equality
$WX = YZ$	Subtraction property of equality

Exercises for Example 2

Use the given information to show the desired statement.

5. Given that $MN = PQ$, show that $MP = NQ$.

6. Given that $AB = DE$ and $BC = CD$, show that $AD = BE$.

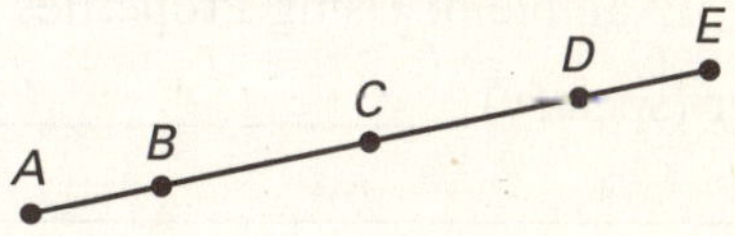

7. Given that $m\angle AQB = m\angle CQD$, show that $m\angle AQC = m\angle BQD$.

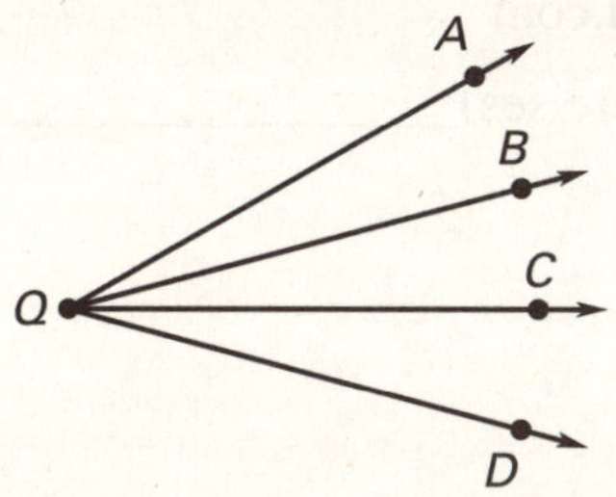

8. Given that $m\angle RPS = m\angle TPV$ and $m\angle TPV = m\angle SPT$, show that $m\angle RPV = 3(m\angle RPS)$

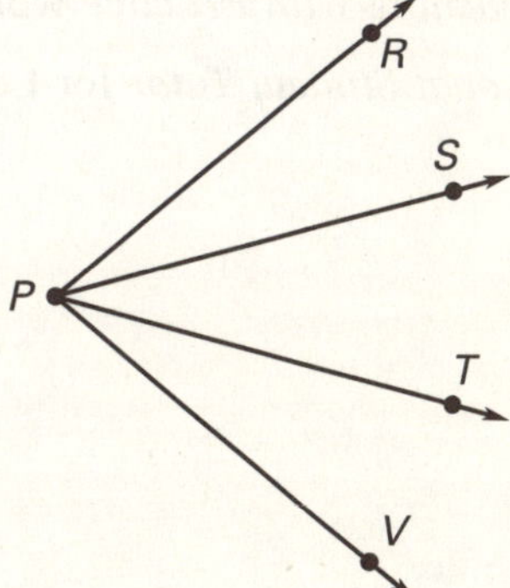

NAME ______________________________ DATE __________

Quick Catch-Up for Absent Students

For use with pages 96–101

The items checked below were covered in class on (date missed) __________

Lesson 2.4: Reasoning with Properties from Algebra

____ **Goal 1:** Use properties from algebra. (pp. 96–97)

Material Covered:

____ Example 1: Writing Reasons

____ Example 2: Writing Reasons

____ Example 3: Using Properties in Real Life

____ **Goal 2:** Use properties of length and measure to justify segment and angle relationships. (p. 98)

Material Covered:

____ Example 4: Using Properties of Length

____ Example 5: Using Properties of Measure

____ Other (specify) ______________________________

____

Homework and Additional Learning Support

____ Textbook (specify) pp. 99–101

____ Internet: Extra Examples at www.mcdougallittell.com

____ *Reteaching with Practice* worksheet (specify exercises) __________

____ *Personal Student Tutor* for Lesson 2.4

LESSON 2.4

NAME ______________________ DATE ________

Cooperative Learning Activity

For use with pages 96–101

GOAL **To understand the use of rules, properties and logic, similar to algebraic properties, to prove theorems and solve problems**

Exploring Algebraic Reasoning

Many algebraic properties, such as the addition, subtraction, and multiplication properties, make it possible to solve problems. Understanding the rules of mathematics is imperative to the learning of these properties. Being able to follow rules in a logical and organized fashion is necessary to fully grasp algebraic concepts. In this activity you will be given a set of rules, similar to properties of algebra, and you will be asked to use these rules and logic to prove a given theorem.

Instructions

Three rules govern the usage of the numbers 1, 2, 3, and 4. A "theorem" is a string of numbers followed by >> followed by a string. By using the three rules, it is possible to change one string of numbers into another string. For example, 1, 2, 2, 3, 4, 4 >> 2, 3, 2 says that by using the rules, 1, 2, 2, 3, 4, 4 can be transformed into 2, 3, 2. Your group will have to prove or disprove a number of challenge strings. The three rules are:

1. If a string ends in the same two numbers, then you may substitute a "2" in place of those two numbers. For example, 3, 4 ,4 >> 3, 2. This is similar to the substitution property.
2. If a string of numbers starts and finishes with the same number, then you may substitute a "3" in place of all the numbers between the first and last numbers. For example 1, 2, 3, 4, 1 >> 1, 3, 1. This is similar to the substitution property.
3. Any two adjacent numbers can change place with each other. For example, 3, 4, >> 4, 3. This is similar to the transitive property.

Example

Show that	1, 2, 2, 3, 4, 4 >> 2, 3, 2	
Proof	1, 2, 2, 3, 4, 4	Given
	1, 2, 2, 3, 2	Rule 1
	2, 1, 2, 3, 2	Rule 3
	2, 3, 2	Rule 2
	So 1, 2, 2, 3, 4, 4 >> 2, 3, 2.	

As a group, use the rules to prove the theorems below.

1. 1, 2, 1, 3, 2 >> 3, 2

2. 1, 2, 2, 2, 2, 1 >> 3, 2

3. 1, 2, 3, 4, 4, 2, 3, >> 3, 2

4. 1, 4, 4, 3, 4, >> 3, 2

5. 1, 4, 3, 2, 2, 3, 4, 2, 1, 4, 4, 4 >> 3, 2

Lesson 2.4

NAME ______________________________ DATE __________

Real-Life Application: When Will I Ever Use This?

For use with pages 96–101

Temperature Scales

The most common temperature scales are Fahrenheit and Celsius. The United States is one of the few countries in the world that does not predominantly use the Celsius scale for everyday temperature measurement. In some cases, both types of measurement are used. Weather reports, for example, usually give a temperature reading in both Celsius and Fahrenheit. Nevertheless, in the United States, the Fahrenheit scale is more commonly used than the Celsius scale.

The Swedish astronomer Anders Celsius developed the Celsius scale in the mid-1700s while the Fahrenheit scale was developed around the same time by a German physicist, Gabriel Daniel Fahrenheit. On a Celsius scale, 0° is the freezing point of water and 100° is its boiling point. This is different on a Fahrenheit scale, where the freezing point of water is 32° and its boiling point is 212°.

In Exercises 1–4, use the following information.

It is sometimes necessary to convert degrees Celsius to degrees Fahrenheit or vice versa. To convert from Celsius to Fahrenheit, you can use the formula

$$F = \frac{9}{5}C + 32$$

where C is the temperature in degrees Celsius and F is the temperature in degrees Fahrenheit.

1. Convert each of the following measurements to degrees Fahrenheit.

a. 20°C **b.** 30°C **c.** −5°C

2. Solve the conversion formula for C and write a reason for each step.

3. Write another equation that is equal to your answer in Exercise 2. What property should you have used to obtain your equation?

4. Convert each of the following measurements to degrees Celsius.

a. 95°F **b.** 14°F **c.** −17.5°F

LESSON **2.4**

NAME ______________________ DATE __________

Challenge: Skills and Applications

For use with pages 96–101

1. **a.** How is the product $4 \cdot 6$ related to 5^2?

   b. How is the product $5 \cdot 7$ related to 6^2?

   c. Make a conjecture about how the product of two positive integers n and $n + 2$ is related to the square of the integer between them.

   d. Write a convincing argument to justify your conjecture.

2. **a.** Find the value of $\frac{1}{1 \cdot 2}$, $\frac{1}{1 \cdot 2} + \frac{1}{2 \cdot 3}$, and $\frac{1}{1 \cdot 2} + \frac{1}{2 \cdot 3} + \frac{1}{3 \cdot 4}$.

   b. Conjecture the value of $\frac{1}{1 \cdot 2} + \frac{1}{2 \cdot 3} + \frac{1}{3 \cdot 4} + \ldots + \frac{1}{n(n+1)}$.

   c. Prove your conjecture. $\left(Hint: \frac{1}{n(n+1)} = \frac{1}{n} - \frac{1}{n+1}\right)$

3. **a.** Prove: The sum of any two consecutive positive integers is an odd number.

   b. Prove: The sum of any three consecutive positive integers is a multiple of 3.

4. **a.** Prove: The square of an even integer is always an even integer.

   b. Prove: The square of an odd integer is always an odd integer.

In Exercises 5–8, decide whether the relationship is *reflexive, symmetric,* and/or *transitive.*

5. **Set:** cities

   Relationship: "is at least as large as"

   Example: Los Angeles is at least as large as Cincinnati.

6. **Set:** angles

   Relationship: "is a supplement of"

   Example: $\angle A$ is a supplement of $\angle B$.

7. **Set:** line segments

   Relationship: "is the same length as"

   Example: $\overline{PQ}$ is the same length as $\overline{RS}$.

8. **Set:** triangles

   Relationship: "has a larger perimeter than"

   Example: $\triangle ABC$ has a larger perimeter than $\triangle DEF$.

LESSON 2.5

TEACHER'S NAME ______________________ CLASS ________ ROOM ______ DATE ______

Lesson Plan

2-day lesson (See *Pacing the Chapter,* TE pages 68C–68D) For use with pages 102–107

GOALS

1. Justify statements about congruent segments.
2. Write reasons for steps in a proof.

State/Local Objectives ____

____

✓ Check the items you wish to use for this lesson.

STARTING OPTIONS

____ Homework Check: TE page 99: Answer Transparencies
____ Warm-Up or Daily Homework Quiz: TE pages 102 and 101, CRB page 68, or Transparencies

TEACHING OPTIONS

____ Motivating the Lesson: TE page 103
____ Lesson Opener (Application): CRB page 69 or Transparencies
____ Technology Activity with Keystrokes: CRB pages 70–72
____ Examples: Day 1: 1–3, SE pages 102–104; Day 2: See the Extra Examples.
____ Extra Examples: Day 1 or Day 2: 1–3, TE pages 103–104 or Transparencies; Internet
____ Closure Question: TE page 104
____ Guided Practice: SE page 104 Day 1: Exs. 1–5; Day 2: See Checkpoint Exs. TE page 104

APPLY/HOMEWORK

Homework Assignment

____ Basic Day 1: 6–11; Day 2: 12–18, 21, 22, 28–40 even
____ Average Day 1: 6–11; Day 2: 12–18, 21, 22, 28–40 even
____ Advanced Day 1: 6–11; Day 2: 12–18, 21–27, 28–40 even

Reteaching the Lesson

____ Practice Masters: CRB pages 73–75 (Level A, Level B, Level C)
____ Reteaching with Practice: CRB pages 76–77 or Practice Workbook with Examples
____ Personal Student Tutor

Extending the Lesson

____ Cooperative Learning Activity: CRB page 79
____ Applications (Interdisciplinary): CRB page 80
____ Challenge: SE page 107; CRB page 81 or Internet

ASSESSMENT OPTIONS

____ Checkpoint Exercises: Day 1 or Day 2: TE pages 104 or Transparencies
____ Daily Homework Quiz (2.5): TE page 107, CRB page 84, or Transparencies
____ Standardized Test Practice: SE page 107; TE page 107; STP Workbook; Transparencies

Notes ____

____

____

LESSON **2.5**

TEACHER'S NAME ________________ CLASS ______ ROOM ______ DATE ______

Lesson Plan for Block Scheduling

1-day lesson (See *Pacing the Chapter,* TE pages 68C–68D) **For use with pages 102–107**

GOALS
1. **Justify statements about congruent segments.**
2. **Write reasons for steps in a proof.**

State/Local Objectives ________________

CHAPTER PACING GUIDE	
Day	**Lesson**
1	Assess Ch. 1; 2.1 (all)
2	2.2 (all)
3	2.3 (all)
4	2.4 (all); **2.5 (begin)**
5	**2.5 (end)**; 2.6 (begin)
6	2.6 (end); Review Ch. 2
7	Assess Ch. 2; 3.1 (all)

✓ **Check the items you wish to use for this lesson.**

STARTING OPTIONS
____ Homework Check: TE page 99: Answer Transparencies
____ Warm-Up or Daily Homework Quiz: TE pages 102 and 101, CRB page 68, or Transparencies

TEACHING OPTIONS
____ Motivating the Lesson: TE page 103
____ Lesson Opener (Application): CRB page 69 or Transparencies
____ Technology Activity with Keystrokes: CRB pages 70–72
____ Examples: Day 4: 1–3, SE pages 102–104; Day 5: See the Extra Examples.
____ Extra Examples: Day 4 or Day 5: 1–3, TE pages 103–104 or Transparencies; Internet
____ Closure Question: TE page 104
____ Guided Practice: SE page 104 Day 4: Exs. 1–5; Day 5: See Checkpoint Exs. TE page 104

APPLY/HOMEWORK
Homework Assignment (See also the assignments for Lessons 2.4 and 2.6.)
____ Block Schedule: Day 4: 6–11; Day 5: 12–18, 21, 22, 28–40 even

Reteaching the Lesson
____ Practice Masters: CRB pages 73–75 (Level A, Level B, Level C)
____ Reteaching with Practice: CRB pages 76–77 or Practice Workbook with Examples
____ Personal Student Tutor

Extending the Lesson
____ Cooperative Learning Activity: CRB page 79
____ Applications (Interdisciplinary): CRB page 80
____ Challenge: SE page 107; CRB page 81 or Internet

ASSESSMENT OPTIONS
____ Checkpoint Exercises: Day 4 or Day 5: TE pages 104 or Transparencies
____ Daily Homework Quiz (2.5): TE page 107, CRB page 84, or Transparencies
____ Standardized Test Practice: SE page 107; TE page 107; STP Workbook; Transparencies

Notes ________________

LESSON 2.5

NAME ________________________ DATE ________

Available as a transparency

WARM-UP EXERCISES

For use before Lesson 2.5, pages 102–107

Give the property that justifies each statement.

1. If $m\angle 1 = m\angle 2$, then $m\angle 2 = m\angle 1$.

2. $m\angle 2 = m\angle 2$

3. If $EF = GH$ and $GH = IJ$ then $EF = IJ$.

4. If $EF = 8$ and $EF = GH$, then $GH = 8$.

DAILY HOMEWORK QUIZ

For use after Lesson 2.4, pages 96–101

Use the property to complete each statement.

1. Multiplication property of equality: If $x = y$, then $ax =$ ______.

2. Symmetric property of equality: If $FG = HJ$, then ______.

3. Transitive property of equality: If $BC = KL$, and $KL = YZ$, then ______.

Solve the equation and state a reason for each step.

4. $3m - 2 = m + 8$

NAME ______________________ DATE __________

Available as a transparency

Application Lesson Opener

For use with pages 102–107

The grading scale at Randy's school is given below.

Percent	90-100	80-89	70-79	60-69	Below 60
Letter Grade	A	B	C	D	F

Teachers can also add a "+" to the letter grade for the highest two percents for any letter grade and a "−" to the letter grade for the lowest two percents.

Randy used the following reasoning to conclude that he earned a B− on his last math test.

I earned an 81% on my last math test. Since 81 is less than 89 but greater than 80, this is between 80 and 89. By definition of a B, I earned a B. Since 80 and 81 are the least two percents, my grade is one of the least two percents that can earn a B. By definition of "−", I earned a B−.

Use Randy's reasoning to complete the two-column proof. For each of Randy's statements listed on the left, his reasoning behind making that statement should be listed on the right.

Statements	**Reasoning**
1. Randy earned an 81% on his last math test.	**1.** Given
2. ______________________	**2.** 81 is less than 89 but greater than 80
3. Randy earned a B on the test.	**3.** ______________________
4. Randy's grade is one of the lowest two percents that can earn a B.	**4.** ______________________
5. Randy earned a B−.	**5.** ______________________

NAME ______________________ DATE ________

Technology Activity

For use with pages 102–107

GOAL Verifying a proof using geometry software

Geometry software can be used to verify a proof. For example, you could use geometry software to construct the diagram described in the proof below. Then, you could use the software's measuring tool to verify the statement about the segment length.

GIVEN $AB = CD$

PROVE $AC = BD$

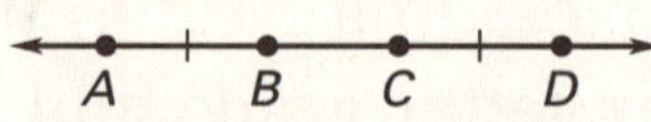

Activity

1. Construct a diagram with points A, B, C, and D such that $AB = CD$ (see figure above).
2. Measure the lengths of $\overline{AB}$, $\overline{CD}$, $\overline{AC}$, and $\overline{BD}$ and verify that $AC = BD$.

Exercises

Use geometry software to verify the following.

1. GIVEN $\overline{AD} \cong \overline{BC}$, $\overline{EC} \cong \overline{ED}$

PROVE $\overline{AE} \cong \overline{BE}$

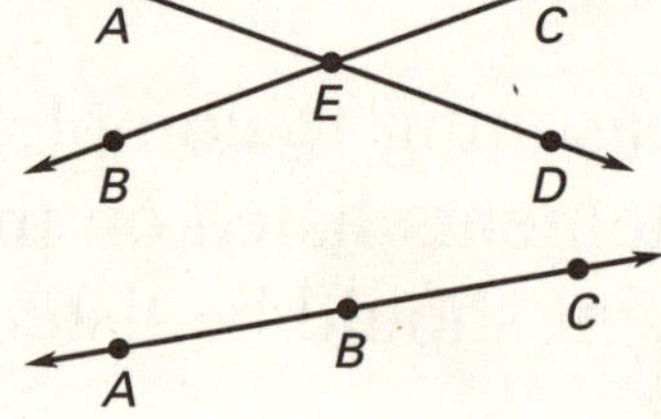

2. GIVEN $AB = BC$

PROVE $AC = 2 \cdot BC$

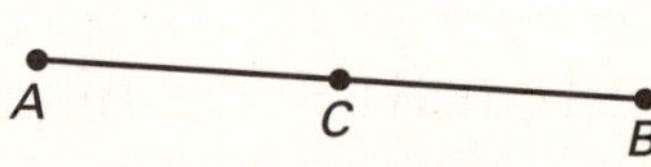

3. GIVEN C is the midpoint of $\overline{AB}$

PROVE $AC = \frac{1}{2}AB$ and $CB = \frac{1}{2}AB$

A C B

NAME _______________ DATE _______

Technology Activity Keystrokes

For use with pages 102–107

TI-92

1. Construct a line with points A, B, C, and D such that $AB = CD$

   F8 9 (Set Coordinate Axes to RECTANGULAR and Grid to ON) ENTER

   F2 5 (Move cursor to point $(-3, 0)$ and prompt says "POINT ON . . .")

   ENTER 2 (Move cursor to point $(-1, 0)$ and prompt says "POINT ON . . .")

   ENTER 2 (Move cursor to point $(1, 0)$ and prompt says "POINT ON . . .")

   ENTER 2 (Move cursor to point $(3, 0)$ and prompt says "POINT ON . . .")

   ENTER 2

   Label the vertices

   F7 4 (Move cursor to point $(-3, 0)$) ENTER A ENTER (Move cursor to point $(-1, 0)$) ENTER B ENTER (Move cursor to point at $(1, 0)$) ENTER C ENTER (Move cursor to point at $(3, 0)$) ENTER D ENTER

   Turn off the axes and the grid

   F8 9 (Set Coordinate Axes and Grid to OFF) ENTER

2. Measure the lengths of $\overline{AB}$, $\overline{CD}$, $\overline{AC}$ and $\overline{BD}$ and verify that $AC = BD$

   F6 1 (Move cursor to A) ENTER (Move cursor to B) ENTER

   (Move cursor to C) ENTER (Move cursor to D) ENTER

   (Move cursor to A) ENTER (Move cursor to C) ENTER

   (Move cursor to B) ENTER (Move cursor to D) ENTER

LESSON 2.5 CONTINUED

NAME ______________________ DATE __________

Technology Activity Keystrokes

For use with pages 102–107

SKETCHPAD

1. Turn on the axes and the grid. Choose **Snap To Grid** from the **Graph** menu.

   Choose the segment straightedge tool. Draw a segment from $(-2.5, 0)$ to $(-0.5, 0)$. Draw a segment from $(-0.5, 0)$ to $(0.5, 0)$. Draw a segment from $(0.5, 0)$ to $(2.5, 0)$.

   Label the points. Choose the text tool. Label point $(-2.5, 0)$ A, label point $(-0.5, 0)$ B, label point $(0.5, 0)$ C, and label point $(2.5, 0)$ D. Turn off the axes and the grid. Choose **Hide Axes** from the **Graph** menu. Choose **Hide Grid** from the **Graph** menu.

2. Measure the lengths of $\overline{AB}$, $\overline{CD}$, $\overline{AC}$, and $\overline{BD}$ and verify that $AC = BD$. Choose the translate selection arrow tool and select $\overline{AB}$. Then hold down the shift key, select $\overline{CD}$, and choose **Length** from the **Measure** menu. To measure the length of $\overline{AC}$, select A, hold down the shift key and select C, and choose **Distance** from the **Measure** menu. To measure the length of $\overline{BD}$, select B, hold down the shift key and select D, and choose **Distance** from the **Measure** menu.

NAME ______________________________ DATE ________

Practice A

For use with pages 102–107

Match the statement with the Property of Congruence.

1. If $\overline{CD} \cong \overline{PM}$ and $\overline{PM} \cong \overline{RV}$, then $\overline{CD} \cong \overline{RV}$.
2. For any segment $\overline{DS}$, $\overline{DS} \cong \overline{DS}$.
3. If $\overline{RA} \cong \overline{DB}$, then $\overline{DB} \cong \overline{RA}$.

A. Symmetric Property
B. Reflexive Property
C. Transitive Property

Mark the diagram with the given information.

4. $\overline{AB} \cong \overline{AC}$

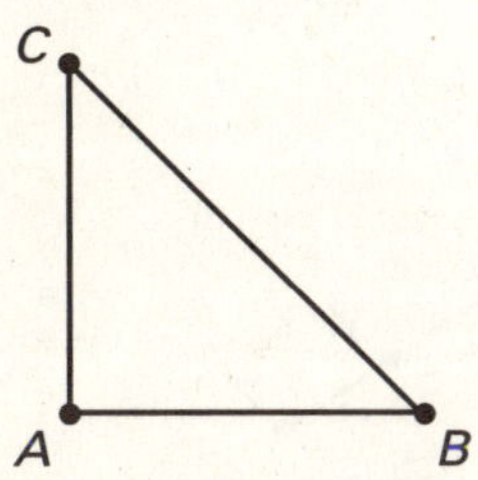

5. $\overleftrightarrow{TR} \perp \overleftrightarrow{MN}$

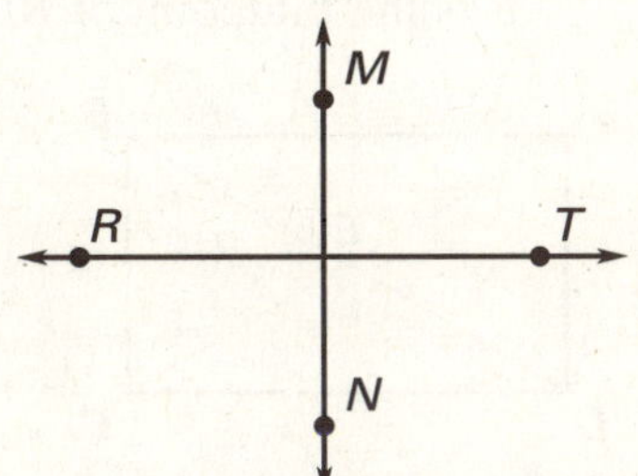

6. $MA = 5, AT = 5$

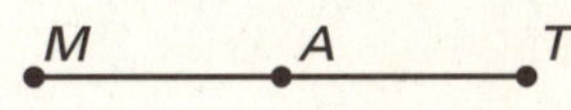

7. $GQ = 4, MQ = 4$
$TQ = 6, PQ = 6$

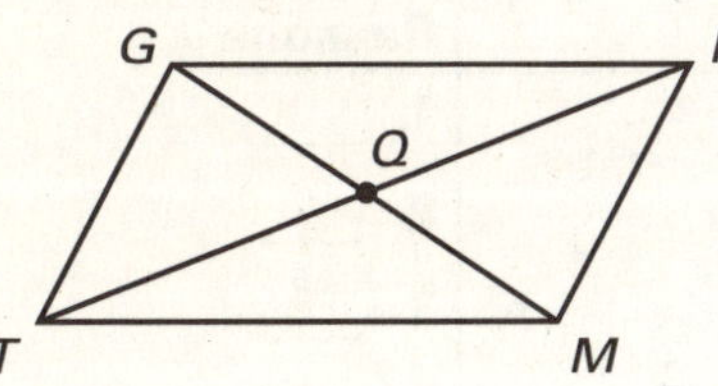

8. A is the midpoint of $\overline{SR}$

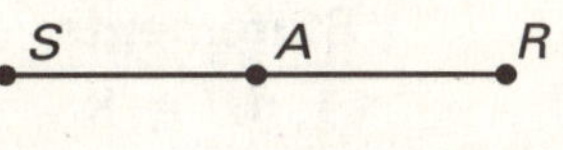

9. $\overline{BN} \cong \overline{DM}$

Complete the argument, giving a reason for each step.

10. **Given:** $\overline{JK} \cong \overline{SH}$, $\overline{SH} \cong \overline{MN}$
Prove: $\overline{JK} \cong \overline{MN}$

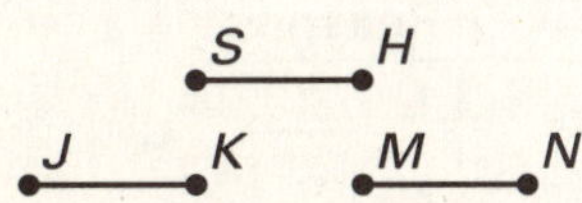

Statements	Reasons
1. $\overline{JK} \cong \overline{SH}$	1. ?
2. $\overline{SH} \cong \overline{MN}$	2. ?
3. $\overline{JK} \cong \overline{MN}$	3. ?

11. **Given:** $\overline{ABCD}$
Prove: $AD = AB + BC + CD$

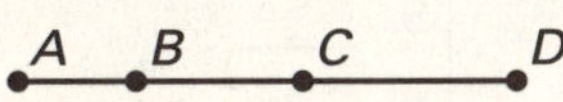

Statements	Reasons
1. $\overline{AD}$	1. Given
2. $AD = AB + BD$	2. ?
3. $BD = BC + CD$	3. ?
4. $AD = AB + BC + CD$	4. ?

12. Write an argument for Exercise 11 in the form of a paragraph proof.

Lesson2.5

NAME ______________________ DATE __________

Practice B

For use with pages 102–107

Match the statement with the Property of Congruence.

1. For any segment $\overline{XY}$, $\overline{XY} \cong \overline{XY}$
2. If $\overline{JK} \cong \overline{MN}$ and $\overline{MN} \cong \overline{CD}$, then $\overline{JK} \cong \overline{CD}$.
3. If $\overline{BN} \cong \overline{TR}$, then $\overline{TR} \cong \overline{BN}$.

A. Transitive Property
B. Symmetric Property
C. Reflexive Property

Mark the diagram with the given information.

4. $LM = 5$, $LP = 5$
   $MN = 3$, $PN = 3$

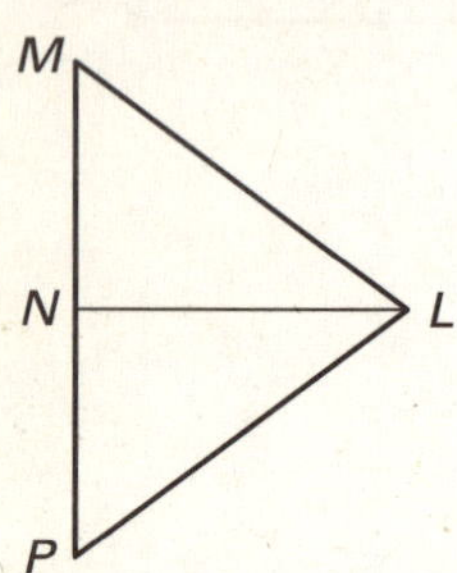

5. E is the midpoint of $\overline{AC}$.
   E is the midpoint of $\overline{BD}$.

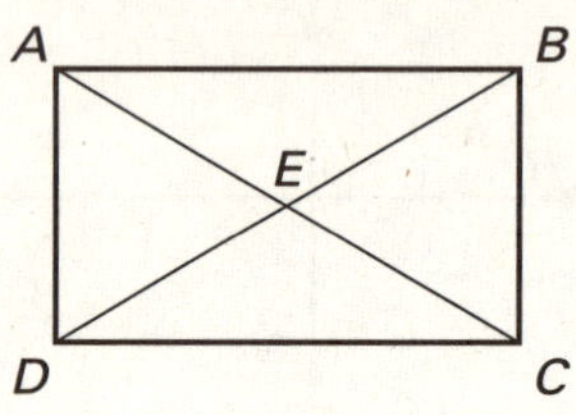

6. $\overline{WX} \cong \overline{YZ}$

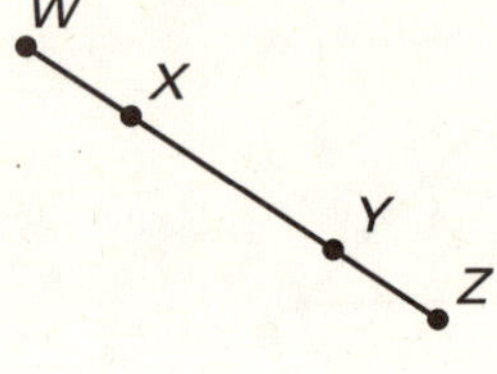

Complete the argument, giving a reason for each step.

7. **Given:** $HI = 8$, $IJ = 8$, $\overline{IJ} \cong \overline{JH}$
   Prove: $\overline{HI} \cong \overline{JH}$

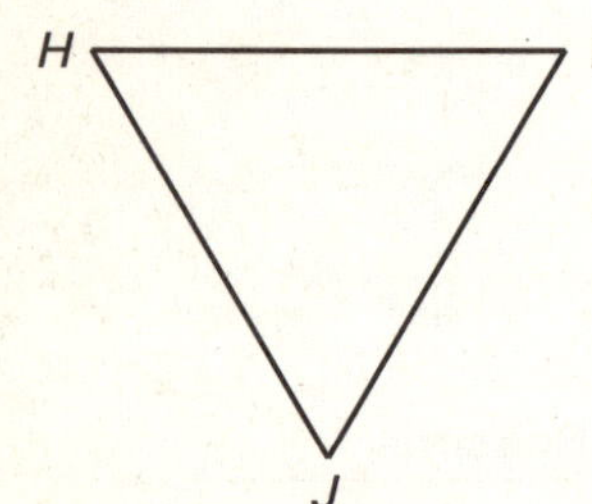

Statements	Reasons
1. $HI = 8$	1. ?
2. $IJ = 8$	2. ?
3. $HI = IJ$	3. ?
4. $\overline{HI} \cong \overline{IJ}$	4. ?
5. $\overline{IJ} \cong \overline{JH}$	5. ?
6. $\overline{HI} \cong \overline{JH}$	6. ?

8. **Given:** $AL = SK$
   Prove: $AS = LK$

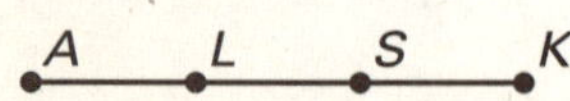

Statements	Reasons
1. $AL = SK$	1. ?
2. $LS = LS$	2. ?
3. $AL + LS = SK + LS$	3. ?
4. $AL + LS = AS$	4. ?
5. $SK + LS = LK$	5. ?
6. $AS = LK$	6. ?

9. Write an argument for Exercise 7 in the form of a paragraph proof.

LESSON 2.5

NAME ______________________________ DATE ________

Practice C

For use with pages 102–107

Lesson 2.5

Mark the diagram with the given information.

1. $GH = 4$, $EF = 4$
   $HE = 6$, $GF = 6$

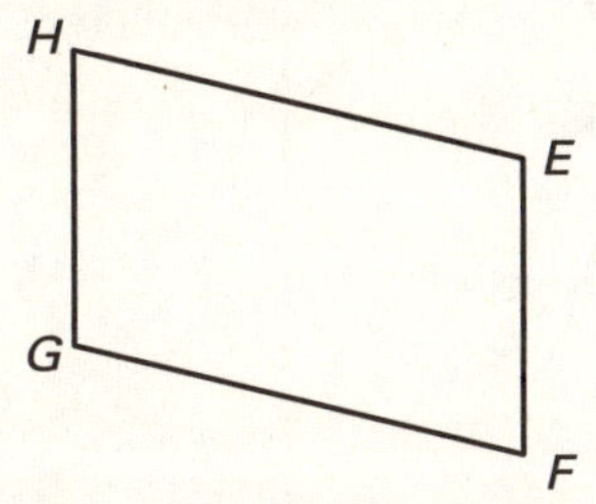

2. N is the midpoint of PQ.

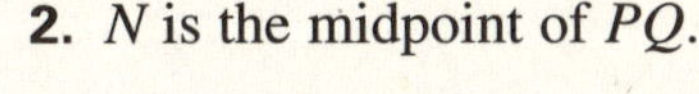

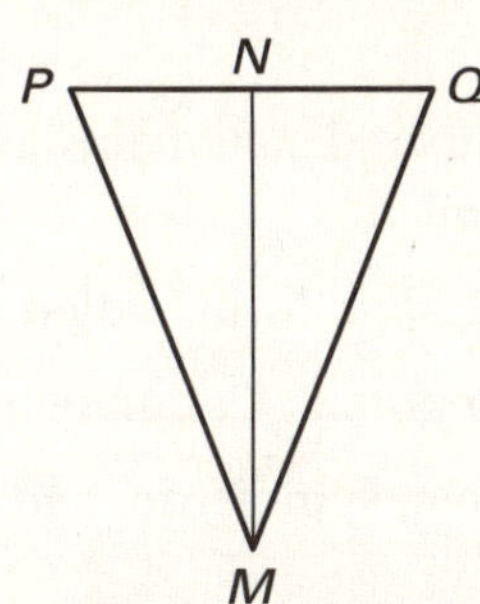

3. $AC = BD$

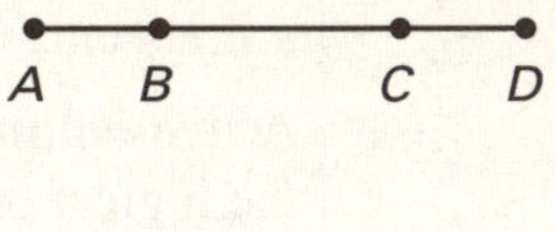

Complete the argument, giving a reason for each step.

4. **Given:** $OP = MN$, $MN = QR$
   Prove: $\overline{OP} \cong \overline{QR}$

M N

O P Q R

Statements	Reasons
1. $OP = MN$, $MN = QR$	1. ?
2. $OP = QR$	2. ?
3. $\overline{OP} \cong \overline{QR}$	3. ?

5. **Given:** $DG = 8$, $GF = 8$, $\overline{GF} \cong \overline{EF}$
   Prove: $\overline{DG} \cong \overline{EF}$

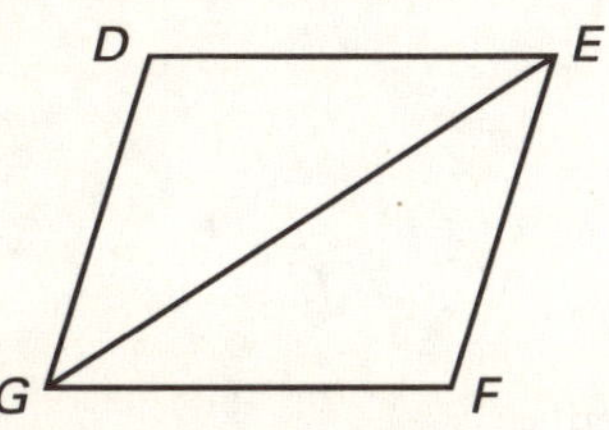

Statements	Reasons
1. $DG = 8$, $GF = 8$	1. ?
2. $DG = GF$	2. ?
3. $\overline{DG} \cong \overline{GF}$	3. ?
4. $\overline{GF} \cong \overline{EF}$	4. ?
5. $\overline{DG} \cong \overline{EF}$	5. ?

6. **Given:** $\overline{FR} \cong \overline{AN}$
   Prove: $\overline{FA} \cong \overline{RN}$

F R A N

Statements	Reasons
1. $\overline{FR} \cong \overline{AN}$	1. ?
2. $FR = AN$	2. ?
3. $RA = RA$	3. ?
4. $FR + RA = AN + RA$	4. ?
5. $FR + RA = FA$	5. ?
6. $AN + RA = RN$	6. ?
7. $FA = RN$	7. ?
8. $\overline{FA} \cong \overline{RN}$	8. ?

7. Write an argument in the form of a paragraph proof.
   Given: T is the midpoint of $\overline{AM}$.
   Prove: $AT = \frac{1}{2}AM$

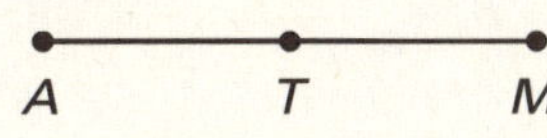

LESSON 2.5

NAME ______________________________ DATE __________

Reteaching with Practice

For use with pages 102–107

GOAL **Write reasons for steps in a proof about segments**

> **VOCABULARY**
>
> A true statement that follows as a result of other true statements is called a **theorem.**
>
> A **two-column proof** has numbered statements and reasons that show the logical order of an argument.
>
> A proof can be written in paragraph form, called a **paragraph** proof.
>
> **Theorem 2.1 Properties of Segment Congruence**
>
> Segment congruence is reflexive, symmetric, and transitive.

EXAMPLE 1 *Using Congruence*

Use the diagram and the given information to complete the missing steps and reasons in the proof.

Given: $AD = 8$, $BC = 8$, $\overline{BC} \cong \overline{CD}$

Prove: $\overline{AD} \cong \overline{CD}$

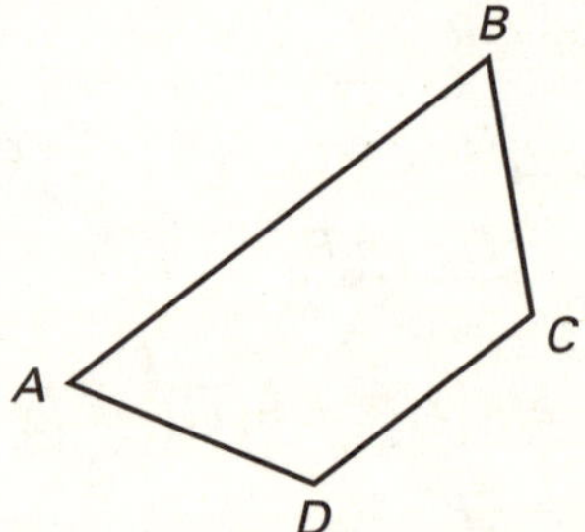

Statements	Reasons
1. a.	**1.** Given
2. $BC = 8$	**2.** b.
3. c.	**3.** Transitive property of equality
4. d.	**4.** Definition of congruent segments
5. $\overline{BC} \cong \overline{CD}$	**5.** e.
6. $\overline{AD} \cong \overline{CD}$	**6.** f.

SOLUTION

a. $AD = 8$ **b.** Given **c.** $AD = BC$ **d.** $\overline{AD} \cong \overline{BC}$

e. Given **f.** Transitive property of congruence

Exercise for Example 1

Use the diagram and the given information to write a proof.

1. Given: $\overline{BC} \cong \overline{CD}$, $\overline{AD} \cong \overline{CD}$,

$AD = 12$, $AB = 12$

Prove: $\overline{BC} \cong \overline{BA}$

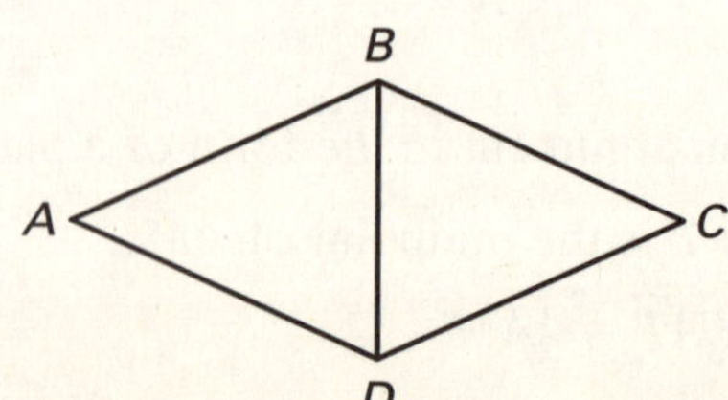

NAME ______________________________ DATE __________

Reteaching with Practice

For use with pages 102–107

EXAMPLE 2 Using Algebra

Solve for the variable using the given information. Explain your steps.

a. $AC = 91$

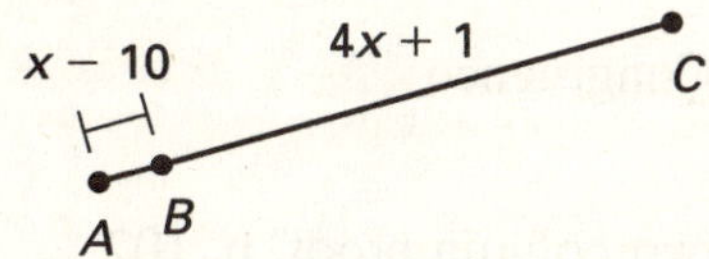

b. $\overline{DE} \cong \overline{EF}, \overline{EF} \cong \overline{GF}$

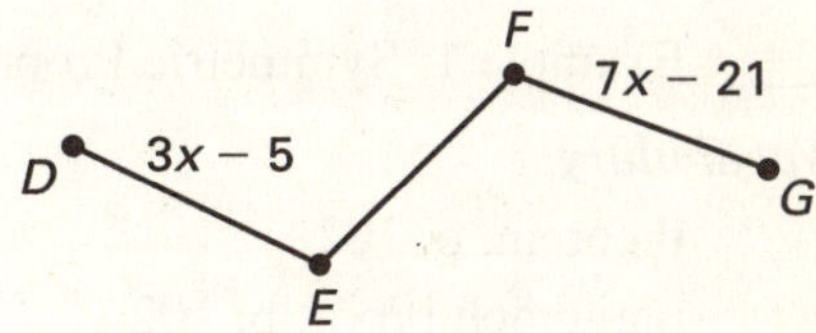

Solution

a. Since the entire segment AC has length 91, the two segments which make up AC must have lengths which add up to 91.

$AB + BC = AC$	Given
$x - 10 + 4x + 1 = 91$	Substitution
$5x - 9 = 91$	Simplify.
$5x = 100$	Addition property of equality
$x = 20$	Division property of equality

b. Because $\overline{DF} \cong \overline{EF}$ and $\overline{EF} \cong \overline{GF}$, $\overline{DE} \cong \overline{GF}$ by the transitive property of congruence. By definition of congruence $DE = GF$.

$3x - 5 = 7x - 21$	Substitution
$3x + 16 = 7x$	Addition property of equality
$16 = 4x$	Subtraction property of equality
$x = 4$	Division property of equality

Exercises for Example 2

Solve for the variable using the given information.

2. Given: $\overline{EG} \cong \overline{HF}$

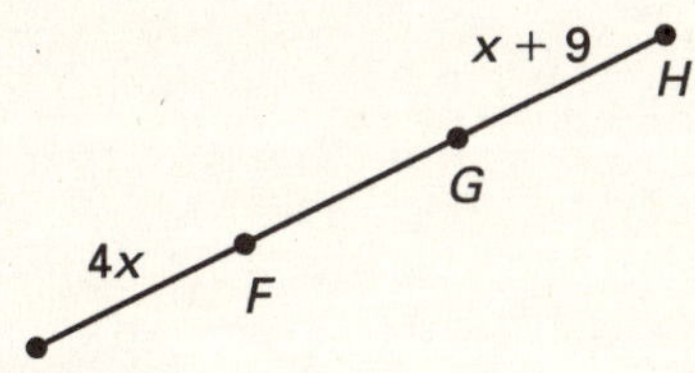

3. Given: $\overline{IP} \cong \overline{JO}, \overline{LM} \cong \overline{IP}$

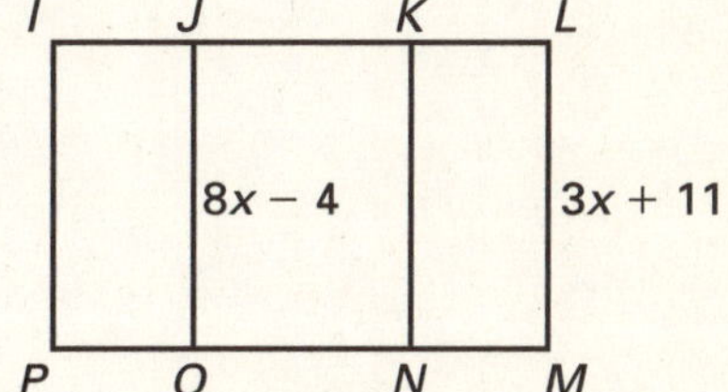

LESSON
2.5

NAME ____________________ DATE ________

Quick Catch-Up for Absent Students

For use with pages 102–107

The items checked below were covered in class on (date missed) __________

Lesson 2.5: Proving Statements about Segments

____ **Goal 1:** Justify statements about congruent segments. (p. 102)

Material Covered:

____ Student Help: Study Tip

____ Example 1: Symmetric Property of Segment Congruence

Vocabulary:

theorem, p. 102 two-column proof, p. 102
paragraph proof, p. 102

____ **Goal 2:** Write reasons for steps in a proof. (pp. 103–104)

Material Covered:

____ Example 2: Using Congruence

____ Student Help: Study Tip

____ Example 3: Using Segment Relationships

____ Activity: Copy a Segment

____ Other (specify) ____________________

Homework and Additional Learning Support

____ Textbook (specify) pp. 104–107

____ *Reteaching with Practice* worksheet (specify exercises) __________

____ *Personal Student Tutor* for Lesson 2.5

NAME ______________________________ DATE ________

Cooperative Learning Activity

For use with pages 102–107

GOAL **To use the properties of segment congruence to understand how the English system of measurement was designed**

Materials: paper, pencil, compass, ruler

Exploring Statements about Segments

Every student has used a ruler to measure distances. In this activity, your group will work with line segments to better understand how the English system of measurement (inches, feet, yards) was formed. The use and proof of statements about segments is instrumental in comprehending the origins of this English system.

Instructions

1. Have one member of your group draw a straight line that is 8 inches long. It is important that this measurement be as precise as possible. Label this segment *AB*.

2. The next member of the group should bisect the line in the fashion shown in Section 1.5 of the textbook. The 8-inch line should now be divided into two line segments that are equal in length. Label the midpoint *C*. Measure $\overline{AC}$ and $\overline{CB}$ to confirm that they are equal in length.

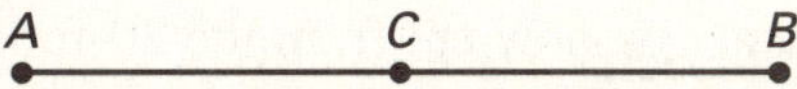

3. Since you know that the above procedure will divide a line into two equal parts, have another group member repeat the process for one of the two equal line segments.
4. Continue with each member of the group repeating this process until the original 8-inch line has been divided into sixteen equal parts.
5. Finally, have a group member measure the sixteen line segments to confirm that they are equal in length.

Analyzing the Results

1. Use a two column proof to prove that $AC = CB$.
2. Use a paragraph proof to prove that $AB = 2AC$.
3. Explain why the English measuring system divides inches into 2, 4, 8, 16 or more equal parts.

NAME ____________________ DATE __________

Interdisciplinary Application

For use with pages 102–107

Concrete Construction

ARCHITECTURE Concrete is one of the most important building materials used today. Office buildings, homes, dams, bridges, highways, and airport runways all use concrete in their construction. In some cases, concrete is sometimes poured around steel rods to help strengthen it. Large structures, such as skyscrapers, always require this type of reinforced concrete, and small structures, such as a residential driveway or a retaining wall, sometimes require this type of concrete as well.

In Exercises 1–4, use the following information.

Suppose you are a construction worker and you need to cut a 20-foot steel rod into individual rods that are 15 inches long. You cut the first 15-inch rod and use it to measure the second rod that is to be cut. After the second rod is cut, you set aside the first rod and use the second rod to measure the third. You follow this pattern for the rest of the cuts.

1. How many 15-inch steel rods can you cut from the 20-foot rod?
2. What property assures you that the length of the first rod you cut will equal the length of the last rod?
3. Suppose you need 22 rods that are 15 inches long. How many cuts will you have to make on another 20-foot steel rod?
4. In Exercise 3, how long is the remaining rod when you are finished?

In Exercises 5–8, use the following information.

Suppose after cutting five rods, you make an error and cut one that is 16 inches.

5. How many 15-inch rods can you now cut from the 20-foot rod?
6. How long is the remaining rod?
7. If you also need to cut 4-inch rods, how many can you cut from the remaining rod?
8. Is there any of the original 20-foot rod left? If so, how many pieces are there and what are their lengths?

LESSON 2.5

NAME ______________________________ DATE ____________

Challenge: Skills and Applications

For use with pages 102–107

In Exercises 1–4, assume that B is the midpoint of $\overline{AC}$, C is the midpoint of $\overline{BD}$, and D is the midpoint of $\overline{BE}$.

1. If $BC = 7$, find DE and AE.
2. If $CE = 9$, find AB and AE.
3. If the coordinate of A is 0 and the coordinate of B is 4, find the coordinates of C, D, and E.
4. If the coordinate of B is 3 and the coordinate of E is -9, find the coordinates of A, C, and D.

In Exercises 5–8, assume that M is the midpoint of $\overline{UV}$, X is the midpoint of $\overline{UM}$, and Y is the midpoint of $\overline{XV}$.

5. If $UV = 16$, find UX and XY.
6. If $MY = 5$, find XY and UV.
7. If the coordinate of X is 7 and the coordinate of Y is 13, find the coordinates of U, M, and V.
8. If the coordinate of U is 2 and the coordinate of M is -10, find the coordinates of X, Y, and V.

In Exercises 9–11, use the diagram shown. Assume that O is the midpoint of $\overline{NP}$, O is the midpoint of $\overline{MQ}$, and $\angle N$ and $\angle P$ are right angles.

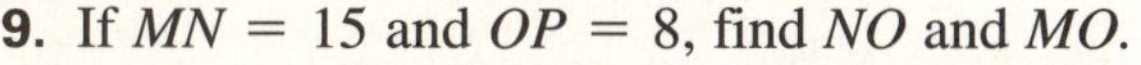

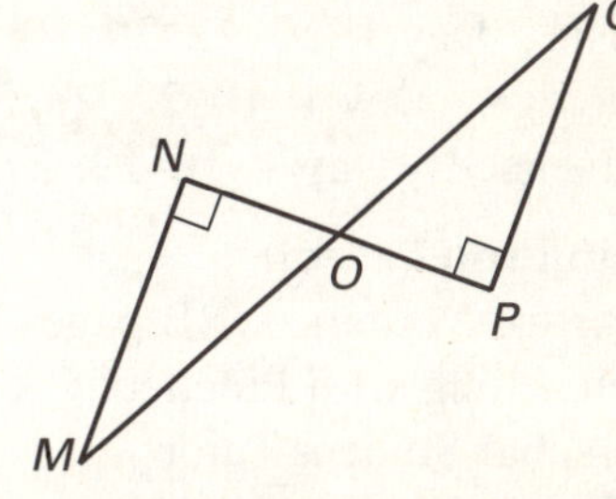

9. If $MN = 15$ and $OP = 8$, find NO and MO.
10. If $OP = 28$ and $PQ = 45$, find MO and MQ.
11. If $MO = 65$ and $NP = 66$, find PQ and MQ.
12. Write a two-column proof. You may use the Pythagorean theorem as a reason.

    Given: $AD = 13$, $CD = x$, $BD = 12$, $\overline{BD} \cong \overline{AC}$, $\angle C$ is a right angle.

    Prove: $x^2 = 25$

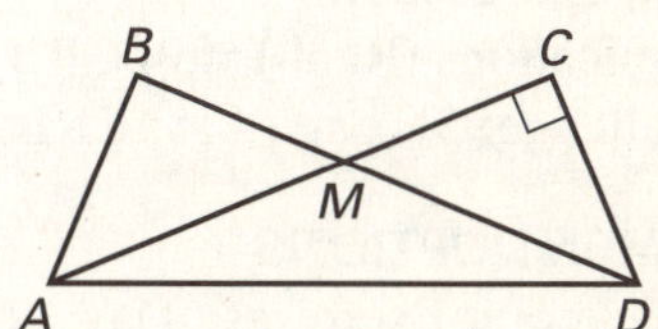

LESSON 2.6

TEACHER'S NAME ______________________ CLASS ________ ROOM ________ DATE ________

Lesson Plan

2-day lesson (See *Pacing the Chapter,* TE pages 68C–68D) **For use with pages 108–116**

GOALS
1. **Use angle congruence properties.**
2. **Prove properties about special pairs of angles.**

State/Local Objectives ____

____

✓ Check the items you wish to use for this lesson.

STARTING OPTIONS

____ Homework Check: TE page 105: Answer Transparencies
____ Warm-Up or Daily Homework Quiz: TE pages 109 and 107, CRB page 84, or Transparencies

TEACHING OPTIONS

____ Concept Activity: SE page 108
____ Lesson Opener (Activity): CRB page 85 or Transparencies
____ Technology Activity with Keystrokes: CRB pages 86–88
____ Examples: Day 1: 1–5, SE pages 109–112; Day 2: 6, SE page 112
____ Extra Examples: Day 1: TE pages 110–112 or Transp.; Day 2: TE page 112 or Transp.; Internet
____ Closure Question: TE page 112
____ Guided Practice: SE page 112 Day 1: Exs. 1, 3, 4, 6, 8; Day 2: Exs. 2, 5, 7, 9

APPLY/HOMEWORK

Homework Assignment

____ Basic Day 1: 10–22, 29, 30; Day 2: 23–28, 33–36, 38–45; Quiz 2: 1–9
____ Average Day 1: 10–22, 29, 30; Day 2: 23–28, 33–36, 38–45; Quiz 2: 1–9
____ Advanced Day 1: 10–22, 29, 30; Day 2: 23–28, 33–45; Quiz 2: 1–9

Reteaching the Lesson

____ Practice Masters: CRB pages 89–91 (Level A, Level B, Level C)
____ Reteaching with Practice: CRB pages 92–93 or Practice Workbook with Examples
____ Personal Student Tutor

Extending the Lesson

____ Applications (Real-Life): CRB page 95
____ Challenge: SE page 115; CRB page 96 or Internet

ASSESSMENT OPTIONS

____ Checkpoint Exercises: Day 1: TE pages 111–112 or Transp.; Day 2: TE page 112 or Transp.
____ Daily Homework Quiz (2.6): TE page 116, or Transparencies
____ Standardized Test Practice: SE page 115; TE page 116; STP Workbook; Transparencies
____ Quiz (2.4–2.6): SE page 116; CRB page 97

Notes ____

____

____

LESSON 2.6

TEACHER'S NAME ______________ CLASS ______ ROOM ______ DATE ______

Lesson Plan for Block Scheduling

1-day lesson (See *Pacing the Chapter,* TE pages 68C–68D) For use with pages 108–116

GOALS
1. **Use angle congruence properties.**
2. **Prove properties about special pairs of angles.**

State/Local Objectives ______________

CHAPTER PACING GUIDE	
Day	**Lesson**
1	Assess Ch. 1; 2.1 (all)
2	2.2 (all)
3	2.3 (all)
4	2.4 (all); 2.5 (begin)
5	2.5 (end); **2.6 (begin)**
6	**2.6 (end)**; Review Ch. 2
7	Assess Ch. 2; 3.1 (all)

✓ Check the items you wish to use for this lesson.

STARTING OPTIONS
____ Homework Check: TE page 105: Answer Transparencies
____ Warm-Up or Daily Homework Quiz: TE pages 109 and 107, CRB page 84, or Transparencies

TEACHING OPTIONS
____ Concept Activity: SE page 108
____ Lesson Opener (Activity): CRB page 85 or Transparencies
____ Technology Activity with Keystrokes: CRB pages 86–88
____ Examples: Day 5: 1–5, SE pages 109–112; Day 6: 6, SE page 112
____ Extra Examples: Day 5: TE pages 110–112 or Transp.; Day 6: TE page 112 or Transp.; Internet
____ Closure Question: TE page 112
____ Guided Practice: SE page 112 Day 5: Exs. 1, 3, 4, 6, 8; Day 6: Exs. 2, 5, 7, 9

APPLY/HOMEWORK
Homework Assignment (See also the assignment for Lesson 2.5.)
____ Block Schedule: Day 5: 10–22, 29, 30; Day 6: 23–28, 33–36, 38–45; Quiz 2: 1–9

Reteaching the Lesson
____ Practice Masters: CRB pages 89–91 (Level A, Level B, Level C)
____ Reteaching with Practice: CRB pages 92–93 or Practice Workbook with Examples
____ Personal Student Tutor

Extending the Lesson
____ Applications (Real-Life): CRB page 95
____ Challenge: SE page 115; CRB page 96 or Internet

ASSESSMENT OPTIONS
____ Checkpoint Exercises: Day 5: TE pages 111–112 or Transp.; Day 6: TE page 112 or Transp.
____ Daily Homework Quiz (2.6): TE page 116, or Transparencies
____ Standardized Test Practice: SE page 115; TE page 116; STP Workbook; Transparencies
____ Quiz (2.4–2.6): SE page 116; CRB page 97

Notes ______________

LESSON
2.6

NAME ______________________________ DATE __________

Available as a transparency

WARM-UP EXERCISES

For use before Lesson 2.6, pages 108–116

Find the measure of each angle.

1. a right angle
2. the complement of 42°
3. the supplement of 42°
4. two congruent angles that are complementary
5. two congruent angles that are supplementary

Lesson 2.6

DAILY HOMEWORK QUIZ

For use after Lesson 2.5, pages 102–107

Solve for the variable using the given information. Explain your steps.

A $2m + 6$ B C $3m - 1$ D

1. Given $\overline{AB} \cong \overline{BC}$, $\overline{CD} \cong \overline{BC}$

Using the segments from Exercises 12–15 on p. 105, along with a straightedge and compass, construct a segment with the given length.

2. $x + y - z$

NAME ______________________ DATE __________

Available as a transparency

Activity Lesson Opener

For use with pages 109–116

SET UP: Work with a partner.

YOU WILL NEED: • scissors

1. Cut out the circle below. Cut the circle along the lines, so that you have five angles.

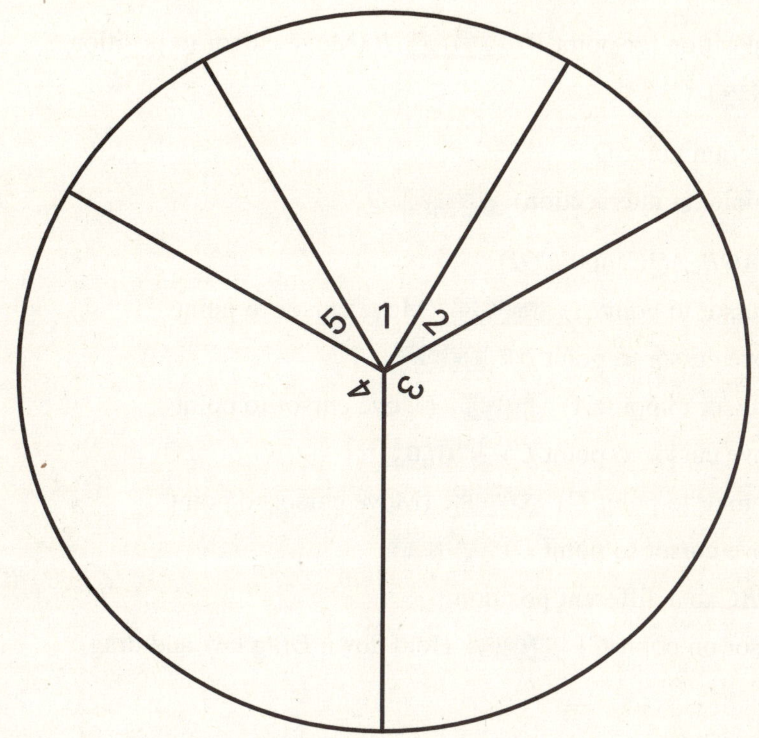

2. Place $\angle 1$ so that its vertex is at the corner of your paper. Use the fact that the corner is a right angle to find an angle that is complementary to $\angle 1$.

3. Find a different angle that is complementary to $\angle 1$. How does this angle compare to the angle you found in Question 2?

4. Place $\angle 1$ so that one of its sides lies along the edge of your paper. Find an angle that is supplementary to $\angle 1$.

5. Find a different angle that is supplementary to $\angle 1$. How does this angle compare to the angle you found in Question 4?

NAME ______________________ DATE __________

Technology Activity Keystrokes

For use with page 110

Keystrokes for Example 3

TI-92

1. Draw intersecting segments *AD* and *BC*

   F2 5 (Locate position for point *A*) ENTER *A* (Move cursor to position for point *D*) ENTER *D*

   F2 5 (Locate position for point *B*) ENTER *B* (Move cursor to position for point *C*) ENTER *C*

2. Construct intersection point *Q*

   F2 3 (Locate point at intersection) ENTER *Q*

3. Measure angles *AQB*, *AQC*, and *CQD*

   F6 3 (Move cursor to point *A*) ENTER (Move cursor to point *Q*) ENTER (Move cursor to point *B*) ENTER

   F6 3 (Move cursor to point *A*) ENTER (Move cursor to point *Q*) ENTER (Move cursor to point *C*) ENTER

   F6 3 (Move cursor to point *C*) ENTER (Move cursor to point *Q*) ENTER (Move cursor to point *D*) ENTER

4. Rotate segment *BC* to a different position

   F1 1 (Put cursor on point *C*) ENTER Hold down Drag key and drag

SKETCHPAD

1. Draw intersecting segments *AD* and *BC*. Select the segment straightedge tool. Draw segments *AD* and *BC* so that they intersect. (Note: You can use the text tool to relabel points.)
2. Construct intersection point *Q*. Select the point tool from the toolbox. Move the cursor to the intersection point of the segments and click. Relabel point *E* to point *Q*.
3. Measure angles *AQB*, *AQC*, and *CQD*. To measure angle *AQB*, choose the translate selection arrow, select point *A*, hold the shift key down, and select points *Q* and *B*. Then select **Angle** from the **Measure** menu. Repeat for the remaining angles. Before selecting the next angle, be sure to click anywhere in the work area to deselect the previous points.
4. Rotate segment *BC* to a different position. Drag point *B*.

NAME ______________________ DATE ________

Technology Activity Keystrokes

For use with page 115

Keystrokes for Exercise 32

TI-92

1. Construct segment *AB*

   F2 5 (Place cursor at desired location for point *A*) ENTER *A* (Move cursor for desired location for point *B*) ENTER *B*

2. Construct a perpendicular line to *AB* through *A*

   F4 1 (Place cursor on segment *AB*) ENTER (Move cursor to point *A*)

   ENTER

3. Construct segment *AC* on perpendicular line

   F2 5 (Move cursor to point *A*) ENTER (Move cursor to a point on the perpendicular line) ENTER *C*

4. Hide perpendicular line

   F7 1 (Place cursor on perpendicular line) ENTER (Line will disappear after next command)

5. Use numerical edit and enter 10

   F7 6 ENTER 10

6. Rotate segment *AB* and *AC* about point *A* 10°

   F5 2 (Select *AB*) ENTER (Select point *A*) ENTER (Select 10) ENTER

   Repeat for *AC*

7. Label these new points *B′* and *C′* F7 4

8. Measure angles *C′AC*, *CAB′*, and *B′AB* F6 3 (Move cursor to point *C′*)

   ENTER (Move cursor to point *A*) ENTER (Move cursor to point *C*) ENTER

   Repeat this process for the other two angles

9. Change the value of the numerical edit

   F1 1 (Place cursor on 10) ENTER ENTER (Delete 10 and type a new number). Repeat as many times as desired

NAME ______________________ DATE __________

Technology Activity Keystrokes

For use with page 115

Keystrokes for Exercise 32

SKETCHPAD

1. Construct segment *AB*. Select the segment straightedge tool and draw segment *AB*.
2. Construct a perpendicular line to *AB* through *A*. Choose the translate selection arrow, select segment *AB*, hold down the shift key and select point *A*. Choose **Perpendicular Line** from the **Construct** menu.
3. Construct segment *AC* on the perpendicular line. Select the segment straightedge tool and draw segment *AC* with point *C* on the perpendicular line.
4. Hide the perpendicular line. Choose the translate selection arrow and select the perpendicular line. Choose **Hide Line** from the **Display** menu.
5. Construct acute angle *EDF* with two segments (make *DF* above *DE*). Select the segment straightedge tool. Draw the angle at a different location on the page than the right angle.
6. Select points *E*, *D*, and *F*, in that order, and mark the angle by selecting **Mark Angle EDF** from the **Transform** menu.
7. Mark *A* as the center by selecting it and choosing **Mark Center A** from the **Transform** menu.
8. Select segments *AB* and *AC*, and points *B* and *C* while holding down the shift key. Rotate $\overline{AB}$ and $\overline{AC}$ about *A* by angle *EDF* by choosing **Rotate** from the **Transform** menu. Click OK.
9. Measure angles *C′AC*, *CAB′*, and *B′AB*. To measure angle *C′AC*, choose the translate selection arrow, hold the shift key down, and select points *C′*, *A*, and *C*. Then select **Angle** from the **Measure** menu. Repeat for the other two angles. Before selecting the next angle, be sure to click anywhere in the work area to deselect the previous points.
10. With the pointer tool selected, drag point *F* and observe the angle measure.

LESSON **2.6**

NAME ______________________ DATE ________

Practice A

For use with pages 109–116

Use the diagram to decide whether the statement is *true* or *false*.

1. If $m\angle 4 = 20°$, then $m\angle 3 = 70°$.
2. If $m\angle 4 = 20°$, then $m\angle 2 = 20°$.
3. $m\angle 4 + m\angle 2 = m\angle 3 + m\angle 1$
4. $m\angle 4 + m\angle 1 = m\angle 3 + m\angle 2$

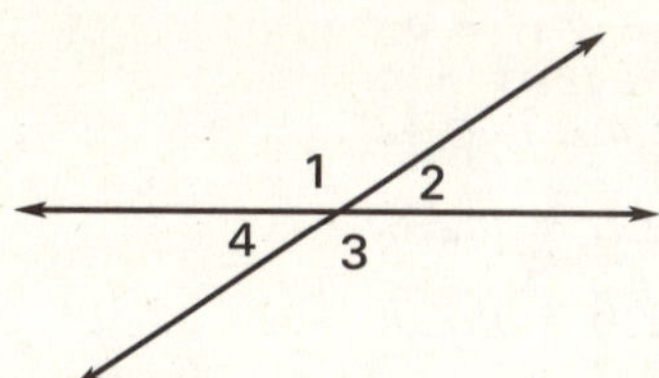

Make a sketch of the given information. Label all angles which can be determined.

5. Vertical angles which measure 40°
6. A linear pair where one angle measures 155°
7. Complementary angles where one angle measures 75°
8. Supplementary angles where one angle measures 75°

Solve for *x*.

9. 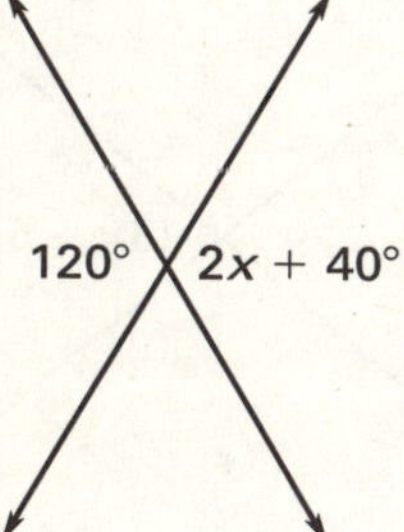

10. 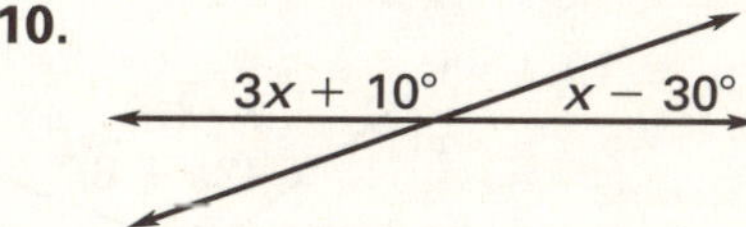

11. 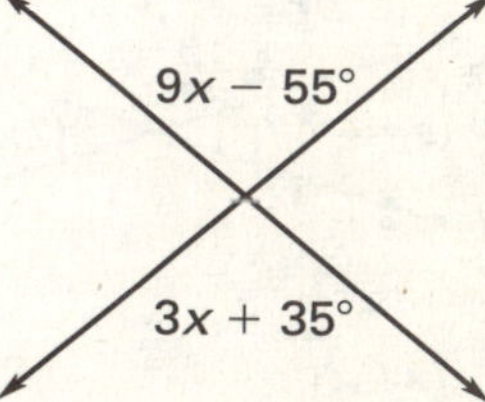

12. Give a reason for each step of the proof. Choose from the list of reasons given.

Given: $\angle 1$ and $\angle 2$ are complementary.
$\angle 1 \cong \angle 3$, $\angle 2 \cong \angle 4$

Prove: $\angle 3$ and $\angle 4$ are complementary.

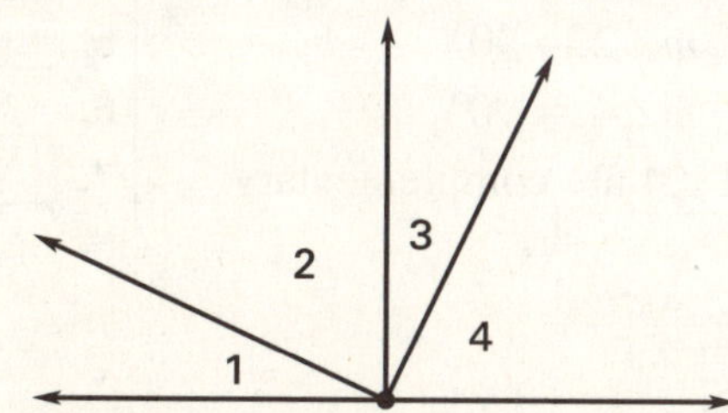

Statements	Reasons
1. $\angle 1$ and $\angle 2$ are complementary.	1. Given
2. $m\angle 1 + m\angle 2 = 90°$	2. ?
3. $\angle 1 \cong \angle 3$, $\angle 2 \cong \angle 4$	3. Given
4. $m\angle 1 = m\angle 3$, $m\angle 2 = m\angle 4$	4. ?
5. $m\angle 3 + m\angle 2 = 90°$	5. ?
6. $m\angle 3 + m\angle 4 = 90°$	6. ?
7. $\angle 3$ and $\angle 4$ are complementary.	7. ?

Reasons: Definition of complementary angles
Definition of congruent angles
Substitution Property of Equality

13. Write a two-column proof.

Given: $\angle 6 \cong \angle 7$

Prove: $\angle 5 \cong \angle 8$

Plan for Proof: First show that $\angle 5 \cong \angle 6$ and $\angle 7 \cong \angle 8$. Then use transitivity to show that $\angle 5 \cong \angle 8$.

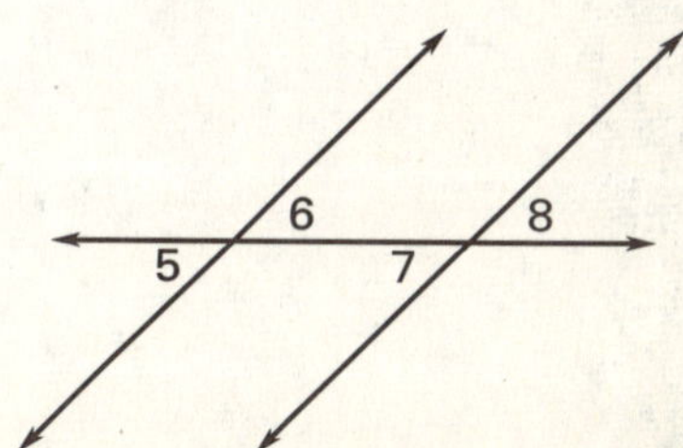

Lesson 2.6

LESSON **2.6**

NAME _______________ DATE _______

Practice B

For use with pages 109–116

Use the diagram to decide whether the statement is *true* or *false*.

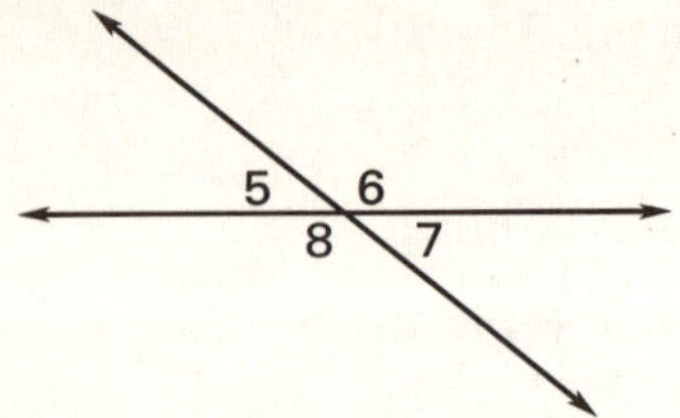

1. If $m\angle 5 = 42°$, then $m\angle 6 = 48°$.
2. If $m\angle 5 = 42°$, then $m\angle 7 = 42°$.
3. $m\angle 5 + m\angle 7 = m\angle 6 + m\angle 8$
4. $m\angle 5 + m\angle 8 = m\angle 6 + m\angle 7$

Lesson 2.6

Make a sketch of the given information. Label all angles which can be determined.

5. Adjacent complementary angles where one angle measures 42°
6. Nonadjacent supplementary angles where where one angle measures 42°
7. Congruent linear pairs
8. Vertical angles which measure 42°

Solve for *x*.

9. (4x − 17)° (2x + 9)°

10.

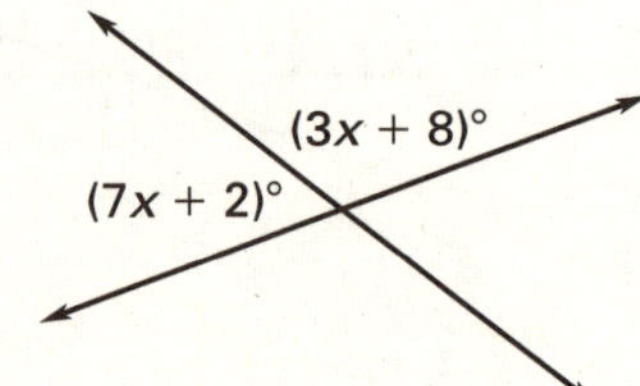

11.

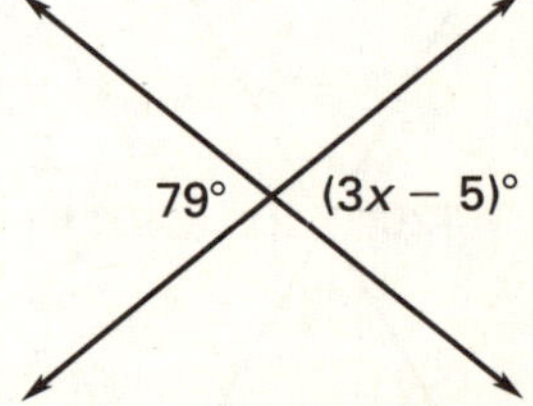

12. Give a reason for each step of the proof.

Given: $\angle 1$ and $\angle 2$ are complementary.
$\angle 1 \cong \angle 3$, $\angle 2 \cong \angle 4$

Prove: $\angle 3$ and $\angle 4$ are complementary.

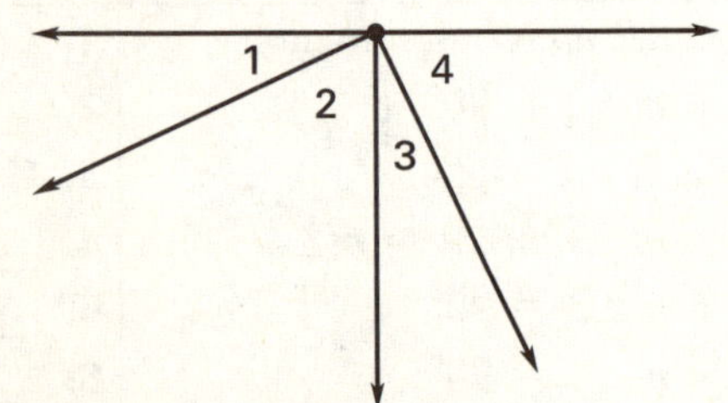

Statements	Reasons
1. $\angle 1$ and $\angle 2$ are complementary.	1. Given
2. $m\angle 1 + m\angle 2 = 90°$	2. ?
3. $\angle 1 \cong \angle 3$, $\angle 2 \cong \angle 4$	3. Given
4. $m\angle 1 = m\angle 3$, $m\angle 2 = m\angle 4$	4. ?
5. $m\angle 3 + m\angle 2 = 90°$	5. ?
6. $m\angle 3 + m\angle 4 = 90°$	6. ?
7. $\angle 3$ and $\angle 4$ are complementary.	7. ?

13. Write a two-column proof.

Given: $\angle 2 \cong \angle 3$

Prove: $\angle 1 \cong \angle 4$

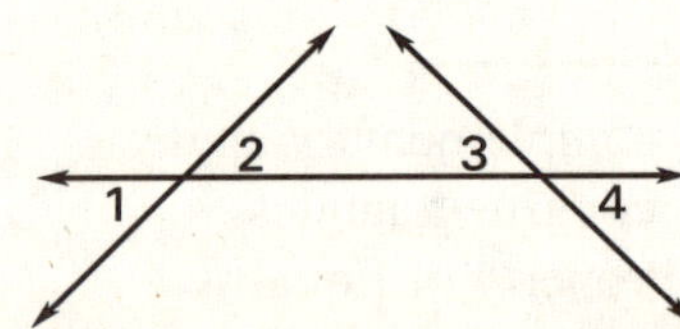

LESSON 2.6

NAME ______________________ DATE ________

Practice C

For use with pages 109–116

Make a sketch of the given information. Label all angles which can be determined.

1. Vertical angles which measure 115°

2. A linear pair where one angle measures 115°

3. Congruent complementary angles

4. Supplementary angles where one angle measures 115°

In Exercises 5–10, complete the statement given that $m\angle BHD = m\angle CHE = m\angle EHF = 90°$.

5. If $m\angle 3 = 42°$, then $m\angle 6 =$ ___?___.

6. If $m\angle BHE = 142°$, then $m\angle 1 =$ ___?___.

7. If $m\angle 1 = 37°$, then $m\angle 6 =$ ___?___.

8. If $m\angle EHG = 132°$, then $\angle 7 =$ ___?___.

9. If $m\angle 7 = 51°$, then $m\angle 3 =$ ___?___.

10. If $m\angle EHB = 153°$, then $m\angle 2 =$ ___?___.

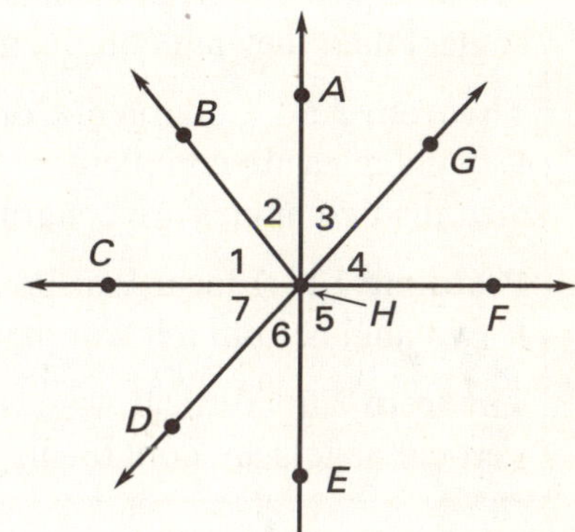

11. Complete the proof.

Given: $\angle 1$ and $\angle 2$ are complementary.
$\angle 1 \cong \angle 3$, $\angle 2 \cong \angle 4$

Prove: $\angle 3$ and $\angle 4$ are complementary.

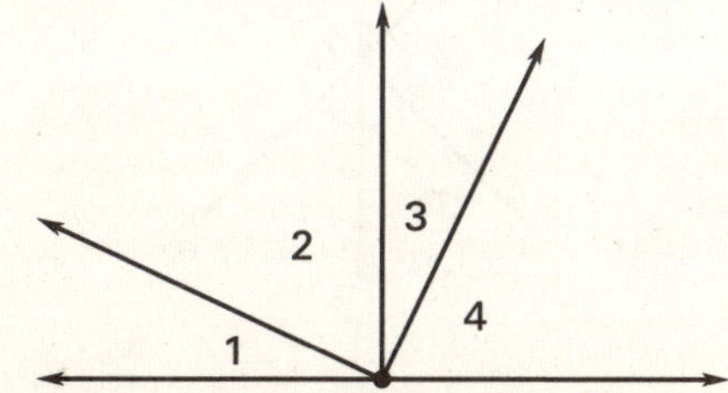

Statements	Reasons
1. ___?___	1. Given
2. $m\angle 1 + m\angle 2 = 90°$	2. ___?___
3. $\angle 1 \cong \angle 3$, $\angle 2 \cong \angle 4$	3. Given
4. ___?___	4. Defintion of congruent angles
5. $m\angle 3 + m\angle 2 = 90°$	5. ___?___
6. $m\angle 3 + m\angle 4 = 90°$	6. ___?___
7. ___?___	7. Definition of complementary angles

12. Write a two-column proof.

Given: $\angle 6 = \angle 7$

Prove: $\angle 5 \cong \angle 8$

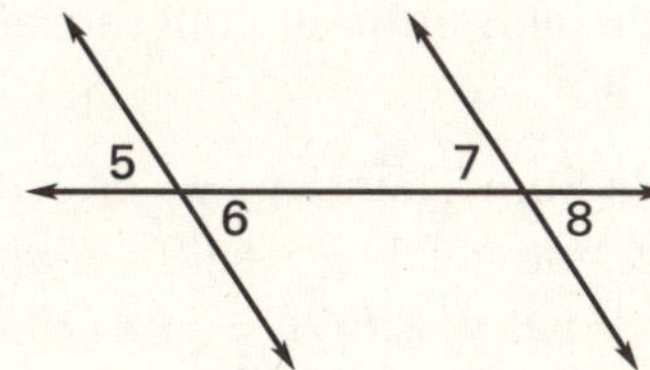

13. Write an argument for Exercise 12 in the form of a paragraph proof.

Lesson 2.6

LESSON 2.6

NAME ______________________ DATE __________

Reteaching with Practice

For use with pages 109–116

GOAL **Use angle congruence properties and prove properties about special pairs of angles**

VOCABULARY

Theorem 2.2 Properties of Angle Congruence
Angle congruence is reflexive, symmetric and transitive.

Theorem 2.3 Right Angle Congruence Theorem
All right angles are congruent.

Theorem 2.4 Congruent Supplements Theorem
If two angles are supplementary to the same angle (or to congruent angles) then they are congruent.

Theorem 2.5 Congruent Complements Theorem
If two angles are complentary to the same angle (or to congruent angles) then the two angles are congruent.

Psotulate 12 Linear Pair Postulate
If two angles form a linear pair, then they are supplementary.

Theorem 2.6 Vertical Angles Theorem
Vertical angels are congruent.

Lesson 2.6

EXAMPLE 1 *Finding Angles*

Complete the statement given that $m\angle AQG = 90°$.

a. $m\angle CQE = ?$

b. If $m\angle BQG = 113°$, then $m\angle EQF = ?$

c. $m\angle AQG + m\angle EQF + m\angle BQC = ?$

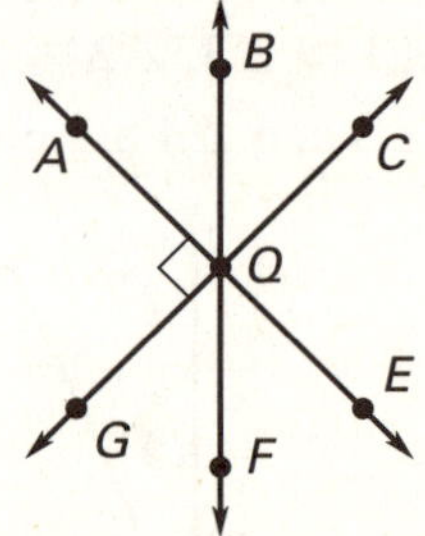

SOLUTION

a. $\angle CQE$ and $\angle AQC$ are vertical angles. By Theorem 2.6, they are congruent. By the definition of congruence, $m\angle CQE = m\angle AQG$, so $m\angle CQE = 90°$.

b. By the Angle Sum Theorem, $m\angle BQG = m\angle AQG + m\angle AQB$. Substituting, you get $113° = 90° + m\angle AQB$, so $m\angle AQB = 23°$ by subtracting. Finally, $\angle EQF \cong \angle AQB$ because they are vertical angles. So, $m\angle EQF = 23°$.

c. $m\angle AQG + m\angle AQB + m\angle BQC = 180°$. $\angle EQF \cong \angle AQB$ because they are vertical angles.
So, $m\angle AQG + m\angle EQF + m\angle BQC = 180°$.

NAME ______________________ DATE __________

Reteaching with Practice

For use with pages 109–116

Exercises for Example 1

Complete the statement given that $m\angle BQD = m\angle CQE = 90°$. Explain your reasoning.

1. $m\angle AQG = ?$

2. $m\angle CQA = ?$

3. If $m\angle CQD = 31°$, then $m\angle EQF = ?$

4. If $m\angle BQG = 125°$, then $m\angle CQF = ?$

5. $m\angle AQB + m\angle GQF + m\angle EQG = ?$

6. If $m\angle EQF = 38°$, then $m\angle BQC = ?$

EXAMPLE 2 Finding Angles

Find the measure of each numbered angle, given that $m\angle DBE = 26°$.

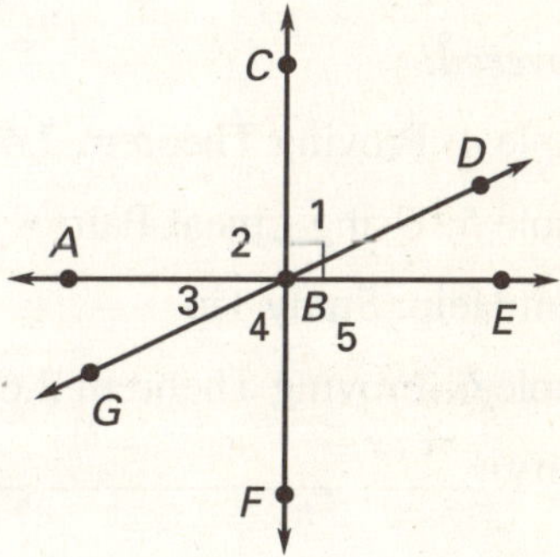

SOLUTION

$\angle CBD$ and $\angle DBE$ are complementary. So, $m\angle 1 = 90° - 26° = 64°$.

$\angle ABC$ and $\angle CBE$ are supplementary. So, $m\angle 2 = 180° - 90° = 90°$.

$\angle ABG$ and $\angle DBE$ are vertical angles. So, $m\angle 3 = 26°$.

$\angle GBF$ and $\angle CBD$ are vertical angles. So, $m\angle 4 = 64°$.

$\angle EBF$ and $\angle CBE$ are supplementary. So, $m\angle 5 = 180° - 90° = 90°$.

Exercises for Example 2

Find the measure of each indicated angle.

7. $\angle ACD$ and $\angle ACB$

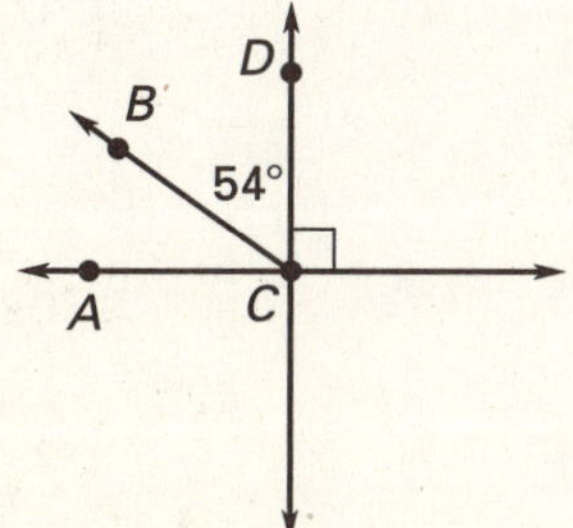

8. $\angle QRT$ and $\angle QRU$

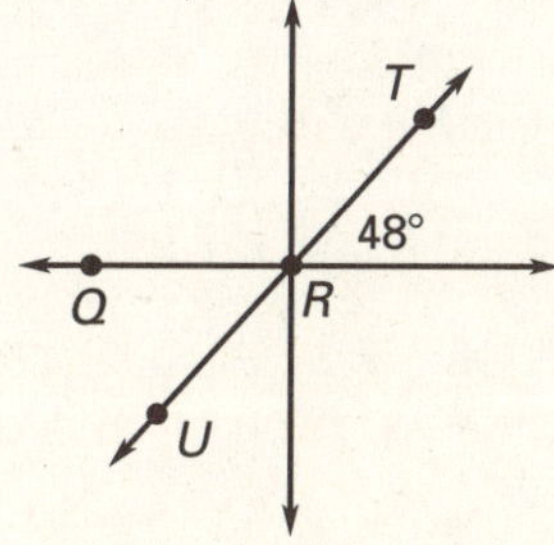

LESSON
2.6

NAME ______________________ DATE ________

Quick Catch-Up for Absent Students

For use with pages 108–116

The items checked below were covered in class on (date missed) ________

Activity 2.6: Investigating Complementary Angles (p. 108)

____ **Goal:** Prove that two angles that are complementary to the same angle are congruent.

____ Student Help: Look Back

Lesson 2.6: Proving Statements about Angles

____ **Goal 1:** Use angle congruence properties. (pp. 109–110)

Material Covered:

____ Example 1: Transitive Property of Angle Congruence

____ Example 2: Using the Transitive Property

____ Example 3: Proving Theorem 2.3

____ Activity: Investigating Supplementary Angles

____ **Goal 2:** Prove properties about special pairs of angles. (pp. 111–112)

Material Covered:

____ Example 4: Proving Theorem 2.4

____ Example 5: Using Linear Pairs

____ Student Help: Study Tip

____ Example 6: Proving Theorem 2.6

____ Other (specify) ______________________

____

Homework and Additional Learning Support

____ Textbook (specify) pp. 112–116

____

____ *Reteaching with Practice* worksheet (specify exercises) ________

____ *Personal Student Tutor* for Lesson 2.6

NAME ______________________ DATE __________

Real-Life Application: When Will I Ever Use This?

For use with pages 109–116

Wallpaper Border

When applying wallpaper border around a window or a door to create a distinctive decorative frame, you need to use congruent angles.

You are wallpapering a border around a window frame. As shown in the diagram at the right, you need to make perfect miter joints at the corners.

The best method to make the corners match would be to make sure that the most prominent design on the border would not be at the corner. Then let the two strips overlap, matching the designs so they look professional. Now take a straightedge and cut through both layers with a knife. Peel away the excess. Do the same on the other corner, making sure your angles are congruent.

In Exercises 1–5, use the diagram.

1. What is the $m\angle 1 + m\angle 3$?
2. What is the $m\angle 2 + m\angle 4$?
3. Do $\angle 1$ and $\angle 3$ have to be equal in measure? $\angle 2$ and $\angle 4$?
4. If $m\angle 3 = 50°$, what is $m\angle 1$?
5. If $m\angle 3 = 52°$, what is $m\angle 4$, $m\angle 1$, and $m\angle 2$?

NAME ______________________________ DATE __________

Challenge: Skills and Applications

For use with pages 109–116

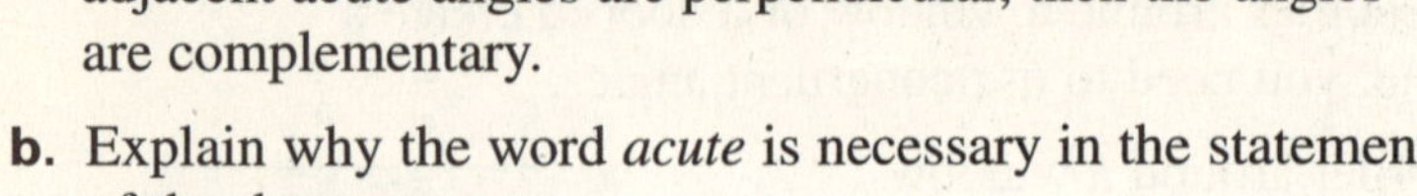

1. **a.** Use the diagram shown to write a two-column proof of the following theorem: If the nonshared sides of two adjacent acute angles are perpendicular, then the angles are complementary.

   b. Explain why the word *acute* is necessary in the statement of the theorem.

   c. Explain how your proof used the fact that the angles are acute.

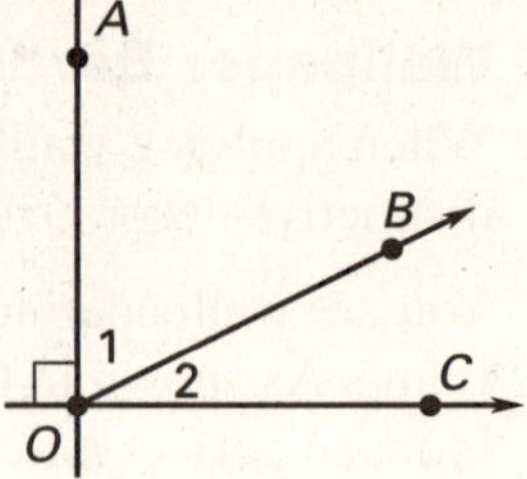

2. Write a two-column proof.

   Given: $\overrightarrow{TV}$ bisects $\angle UTW$

   Prove: $\overrightarrow{TY}$ bisects $\angle XTZ$

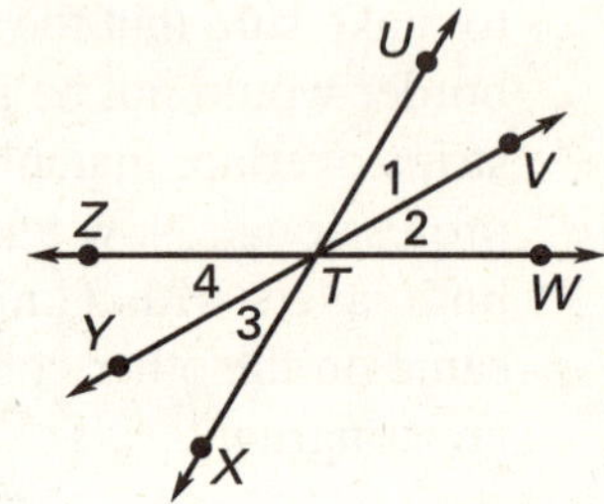

3. Write a paragraph proof to show that if two lines form congruent adjacent angles, then the lines are perpendicular.

   Given: $\angle 1 \cong \angle 2$

   Prove: $\overleftrightarrow{KM} \perp \overleftrightarrow{LN}$

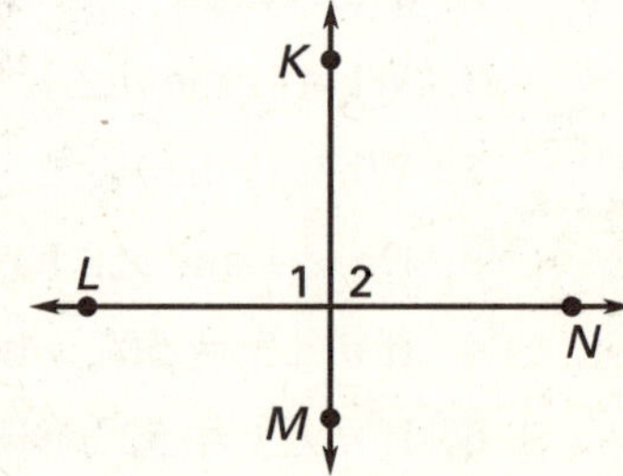

4. Write a paragraph proof. You may use the result of Exercise 3.

   Given: $\overrightarrow{MK}$ bisects $\angle PMQ$; $\angle 1 \cong \angle 4$

   Prove: $\angle 1$ and $\angle 2$ are complementary.

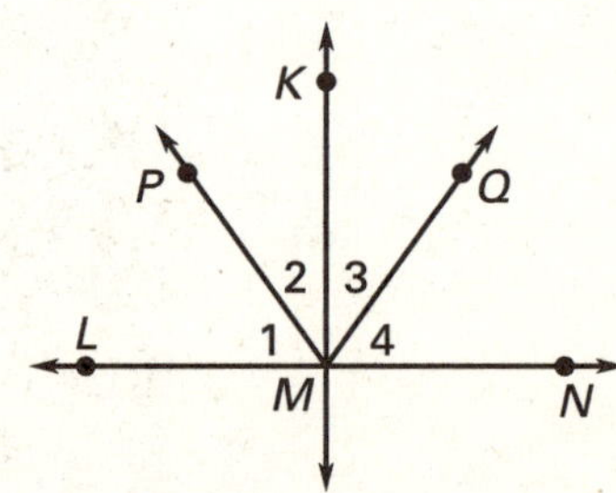

5. In the diagram, $\overrightarrow{OB}$ bisects $\angle AOC$, and $\overrightarrow{OD}$ bisects $\angle COE$.

   a. Make a conjecture about the relationship between $\overrightarrow{OB}$ and $\overrightarrow{OD}$.

   b. Write a two-column proof that your conjecture is correct.

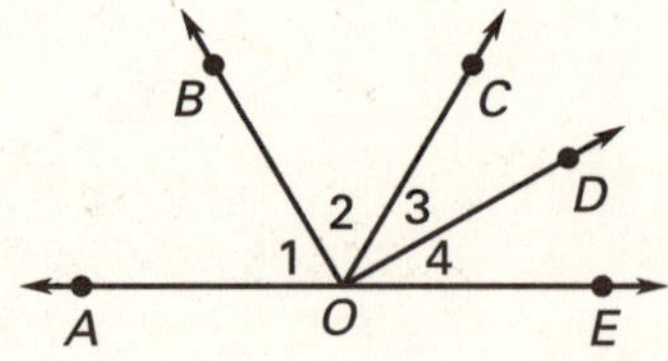

NAME ______________________ DATE __________

Quiz 2

For use after Lessons 2.4–2.6

Solve the equation and state a reason for each step. *(Lesson 2.4)*

1. $x + 5 = 7$
2. $x - 6 = -3$
3. $3x + 4 = 13$
4. $x + 5 = 2x + 5$
5. $2(4x + 2) = 20$
6. $-(-3x + 2) = 13$

Proof **In Exercises 7 and 8, write a two column proof.** *(Lesson 2.5 and Lesson 2.6)*

7. **Given:** $\angle AXB \cong \angle CXD$

   Prove: $\angle AXC \cong \angle BXD$

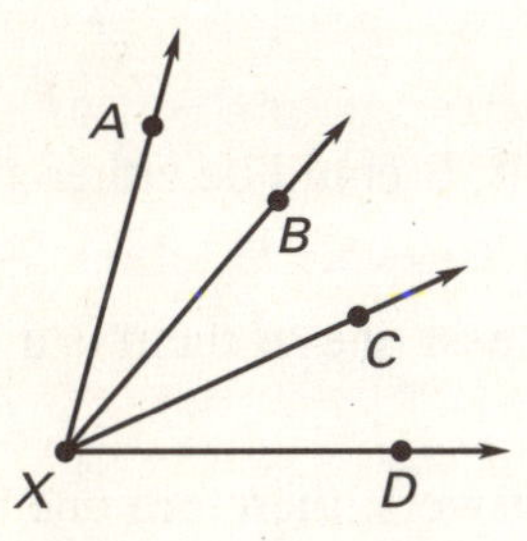

Statements	Reasons
1. ________	1. ________
2. ________	2. ________
3. ________	3. ________
4. ________	4. ________
5. ________	5. ________
6. ________	6. ________
7. ________	7. ________
8. ________	8. ________

8. **Given:** $\overline{XY} \cong \overline{ZW}$

   Prove: $\overline{XZ} \cong \overline{YW}$

X Y Z W

Statements	Reasons
1. ________	1. ________
2. ________	2. ________
3. ________	3. ________
4. ________	4. ________
5. ________	5. ________
6. ________	6. ________
7. ________	7. ________
8. ________	8. ________

9. ***Fireworks*** You are watching fireworks. Use the diagram below to find the angle of inclination.

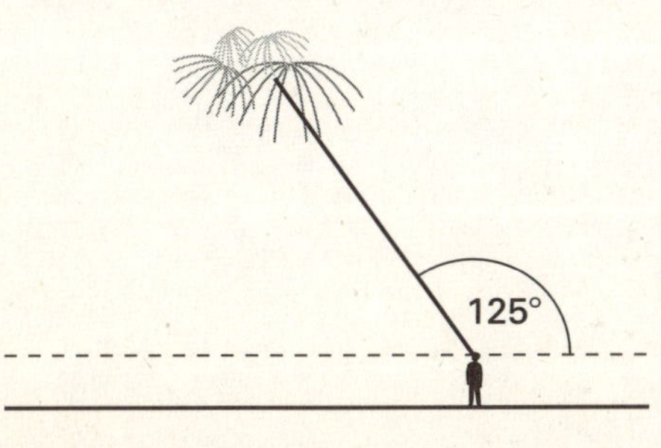

Answers

1. ________
2. ________
3. ________
4. ________
5. ________
6. ________
7. See left.
8. See left.
9. ________

NAME ______________________________ DATE __________

Chapter Review Games and Activities

For use after Chapter 2

The following is a fun problem that will help you with your logic and reasoning skills. To answer the question you need to use "if-then" arguments.

You are visiting a strange land in which every person is either a knight or a knave. Knights always tell the truth, and knaves always lie. You want to visit the castle. You come to a place in the road where you must turn either right or left. Standing there are two people, person A and person B. You say to them "Tell me about yourselves." Person A says "At least one of us is a knight." Then you say to them "Which way should I go to reach the castle?" Person A says, "Go left." Person B says, "Go right." Which way should you go?

Solution

First, let us investigate the answers to your first question.

Suppose that A is telling the truth. If A is telling the truth, then A must be a knight. B could be either a knight or a knave.

Suppose A is telling a lie. If A is telling a lie, then it must not be the case that at least one of them is a knight. Hence, they must both be knaves.

Now let us investigate the answer to your second question. They gave different answers, therefore one of them must be a knight and one of them must be a knave. From our reasoning above, it must be the case that A is a knight and B is a knave. Therefore you should go left.

Review and Assess

NAME ______________________________ DATE __________

Chapter Test A

For use after Chapter 2

Write the inverse, converse, and contrapositive of the conditional statement.

1. If the sky is clear, then we can see the stars at night.

   Inverse: ______________________________

   Converse: ______________________________

   Contrapositive: ______________________________

2. If I go to the beach, then I will go swimming.

   Inverse: ______________________________

   Converse: ______________________________

   Contrapositive: ______________________________

3. If the point is on the line, then it is in the plane.

   Inverse: ______________________________

   Converse: ______________________________

   Contrapositive: ______________________________

4. If we use a calculator, then we are in math class.

   Inverse: ______________________________

   Converse: ______________________________

   Contrapositive: ______________________________

Determine whether the bicondidional statement is *true* or *false*.

5. The milk is chocolate if and only if the cow is brown.
6. The figure is a triangle if and only if it has three sides.

Using *p* and *q* below, write the symbolic statement.

p: The angle has a measure of 90°.

q: The angle is a right angle.

7. $p \to q$
8. $\sim q$
9. $\sim p$
10. $q \to p$
11. $\sim p \to \sim q$
12. $\sim q \to \sim p$

Answers

1. See left. __________
2. See left. __________
3. See left. __________
4. See left. __________
5. __________
6. __________
7. __________
8. __________
9. __________
10. __________
11. __________
12. __________

NAME ______________________ DATE __________

Chapter Test A

For use after Chapter 2

Name the property used to make the conclusion.

13. If $x + 7 = 23$, then $x = 16$.

14. If $4x = 40$, then $x = 10$.

15. If $x = 7$ and $y = 7$, then $x = y$.

16. If $4x = 28$, then $28 = 4x$.

17. If $x - 24 = 25$, then $x = 49$.

Solve for the variable using the given information.

18. Given: $GM = 28$

x + 3   2x − 8
G   A   M

19. Given: $\overline{AB} \cong \overline{BC}$, $\overline{BC} \cong \overline{CD}$

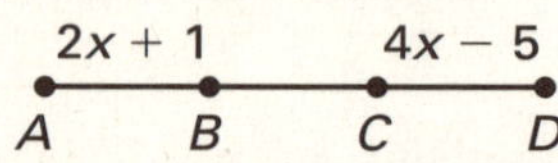

Use the diagram and given information to complete the missing reasons in the proof.

20. *Given:* $\angle 1$ and $\angle 2$ are vertical angles.

Prove: $m\angle 1 = m\angle 2$

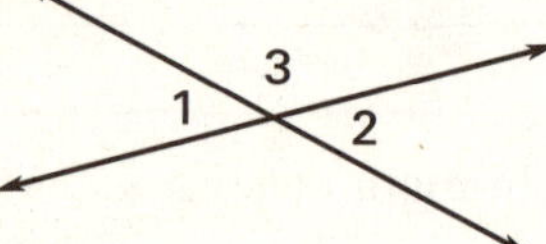

Statements	Reasons
1. $m\angle 1 + m\angle 3 = 180°$ $m\angle 2 + m\angle 3 = 180°$	1. ________
2. $m\angle 1 + m\angle 3 = m\angle 2 + m\angle 3$	2. ________
3. $m\angle 1 = m\angle 2$	3. ________

21. *Given:* $AB = BC$

Prove: $\frac{1}{2}AC = BC$

A   B   C

Statements	Reasons
1. $AB = BC$	1. ________
2. $AC = AB + BC$	2. ________
3. $AC = BC + BC$	3. ________
4. $AC = 2BC$	4. ________
5. $\frac{1}{2}AC = BC$	5. ________

Answers

13. ________
14. ________
15. ________
16. ________
17. ________
18. ________
19. ________
20. See left.
21. See left.

CHAPTER 2

NAME ______________________ DATE __________

Chapter Test B

For use after Chapter 2

Write the inverse, converse, and contrapositive of the conditional statement.

1. If the sky is clear, then we can see the stars at night.

Inverse: ______________________

Converse: ______________________

Contrapositive: ______________________

2. If I ride on the Ferris wheel, then I am not afraid of heights.

Inverse: ______________________

Converse: ______________________

Contrapositive: ______________________

3. If the point is on the line, then it is in the plane.

Inverse: ______________________

Converse: ______________________

Contrapositive: ______________________

4. If we use a calculator, then we are in math class.

Inverse: ______________________

Converse: ______________________

Contrapositive: ______________________

Determine whether the bicondidional statement about the diagram is *true* or *false*.

5. $\angle 1$ and $\angle 2$ are supplementary if and only if they form a linear pair.

6. $\angle 3$ and $\angle 2$ are equal if and only if $\angle 1$ and $\angle 3$ are equal.

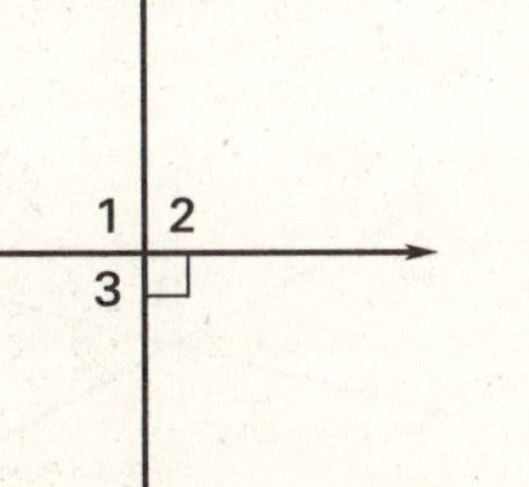

Using *p* and *q* below, write the symbolic statement in words.

p: Points *X*, *Y*, and *Z* are noncollinear.

q: Points *X*, *Y*, and *Z* are coplanar.

7. $p \rightarrow q$ **8.** $\sim q$ **9.** $\sim p$

10. $q \rightarrow p$ **11.** $\sim p \rightarrow \sim q$ **12.** $\sim q \rightarrow \sim p$

Answers

1. See left.
2. See left.
3. See left.
4. See left.
5. ______
6. ______
7. ______
8. ______
9. ______
10. ______
11. ______
12. ______

Review and Assess

NAME ______________________ DATE __________

Chapter Test B

For use after Chapter 2

Name the property used to make the conclusion.

13. If $x + 7 = 23$, then $x = 16$.

14. If $4x = 40$, then $x = 10$.

15. If $x = 7$ and $y = 7$, then $x = y$.

16. If $4x = 28$, then $28 = 4x$.

17. If $x - 24 = 25$, then $x = 49$.

Solve for the variable using the given information.

18. Given: $GM = 28$

$4x + 9$ $3x - 2$

G   E   M

19. Given: $\overline{AB} \cong \overline{BC}$, $\overline{BC} \cong \overline{CD}$

$5x - 8$ $3x + 20$

A   B   C   D

Write a two-column proof.

20. *Given:* $AB = BC$

Prove: $\frac{1}{2}AC = BC$

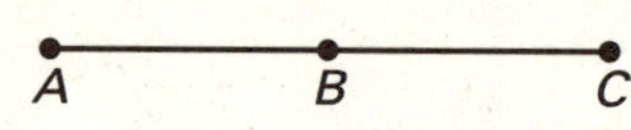

Statements	Reasons
1. ________	1. ________
2. ________	2. ________
3. ________	3. ________
4. ________	4. ________
5. ________	5. ________

21. *Given:* $\angle 1$ and $\angle 2$ are vertical angles.

Prove: $m\angle 1 = m\angle 2$

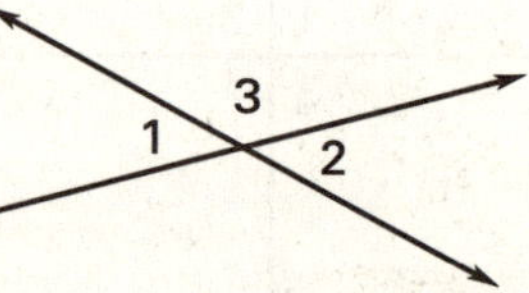

Statements	Reasons
1. ________	1. ________
________	________
2. ________	2. ________
________	________
3. ________	3. ________
________	________

Answers

13. ________

14. ________

15. ________

16. ________

17. ________

18. ________

19. ________

20. See left.

21. See left.

CHAPTER 2

NAME ______________________ DATE __________

Chapter Test C

For use after Chapter 2

Write the inverse, converse, and contrapositive of the conditional statement.

1. If the sky is clear, then we can see the stars at night.

   Inverse: ______________________

   Converse: ______________________

   Contrapositive: ______________________

2. If I sail on a boat, then I will get seasick.

   Inverse: ______________________

   Converse: ______________________

   Contrapositive: ______________________

3. If the angle is less than 90°, then it is an acute angle.

   Inverse: ______________________

   Converse: ______________________

   Contrapositive: ______________________

4. If the tree is an evergreen, then it does not lose its leaves in the fall.

   Inverse: ______________________

   Converse: ______________________

   Contrapositive: ______________________

Rewrite the biconditional statement as a conditional statement.

5. Two segments are congruent if and only if they have the same length.

6. An angle is a right angle if and only if it has a measure of 90°.

Using *p* and *q* below, write the symbolic statement.

p: Points X, Y, and Z are noncollinear.

q: Points X, Y, and Z form a triangle.

7. $p \to q$
8. $\sim q$
9. $\sim p$
10. $q \to p$
11. $\sim p \to \sim q$
12. $\sim q \to \sim p$

Answers

1. See left.
2. See left.
3. See left.
4. See left.
5. ______________
6. ______________
7. ______________
8. ______________
9. ______________
10. ______________
11. ______________
12. ______________

Review and Assess

CHAPTER 2 CONTINUED

NAME ____________________ DATE ________

Chapter Test C

For use after Chapter 2

Name the property used to make the conclusion.

13. If $x + 7 = 23$, then $x = 16$.

14. If $4x = 40$, then $x = 10$.

15. If $x = 7$ and $y = 7$, then $x = y$.

16. If $4x = 28$, then $28 = 4x$.

17. If $x - 24 = 25$, then $x = 49$.

Solve for the variable using the given information.

18. Given: $GM = 28$

G, E, M; $3x + 9$ (GE), $8x - 25$ (EM)

19. Given: $\overline{AB} \cong \overline{BC}$, $\overline{BC} \cong \overline{CD}$

A, B, C, D; $\frac{x}{2} + 5$ (AB), $3x - 10$ (CD)

Write a two-column proof.

20. Given: $\overline{AB} \cong \overline{CD}$

$\overline{CD} \cong \overline{EF}$

Prove: $\overline{AB} \cong \overline{EF}$

Statements	Reasons
1. ________	1. ________
2. ________	2. ________
3. ________	3. ________
4. ________	4. ________

21. Given: $\angle 2 \cong \angle 3$

$\angle 4 \cong \angle 5$

Prove: $\angle 1$ is supplementary to $\angle 6$

1, 2, 3, 4, 5, 6

Statements	Reasons
1. ________	1. ________
2. ________	2. ________
3. ________	3. ________
4. ________	4. ________
5. ________	5. ________
6. ________	6. ________

Answers

13. ________

14. ________

15. ________

16. ________

17. ________

18. ________

19. ________

20. See left.

21. See left.

NAME ______________________________ DATE __________

SAT/ACT Chapter Test

For use after Chapter 2

1. What is the converse of "If it is raining, then we will not go to the beach?"

Ⓐ If it is not raining, then we will go to the beach.

Ⓑ We will go to the beach if and only if it is not raining.

Ⓒ If we go to the beach, then it is not raining.

Ⓓ If we do not go to the beach, then it is raining.

Ⓔ None of the above

2. "If $AB = 10$, then $10 + BC = 24$" is an example of the

Ⓐ Symmetric property of equality.

Ⓑ Substitution property of equality.

Ⓒ Addition property of equality.

Ⓓ Subtraction property of equality.

Ⓔ Reflexive property of equality.

3. Two angles $\angle 3$ and $\angle 5$ are supplementary to $\angle 4$. If $m\angle 5$ is 78°, what is $m\angle 3$?

Ⓐ 102° Ⓑ 2°

Ⓒ 78° Ⓓ 90°

Ⓔ 180°

4. In $ABCD$, $\overline{AB} \cong \overline{CD}$ and $\overline{BC} \cong \overline{AD}$. What is the value of x?

Ⓐ 5

Ⓑ 10

Ⓒ 7

Ⓓ 6

Ⓔ 15

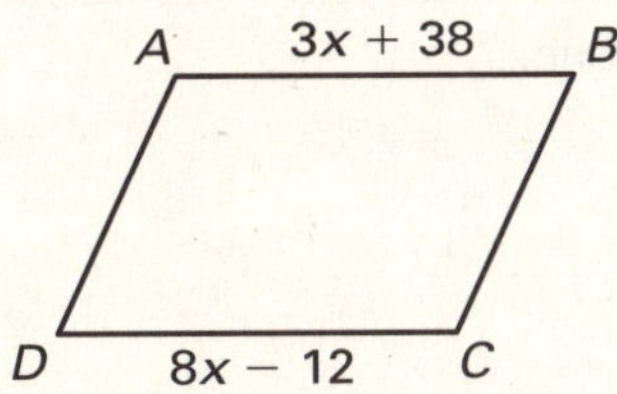

5. In the diagram, $\angle 1 \cong \angle 2$. Which of the folowing is *not* true?

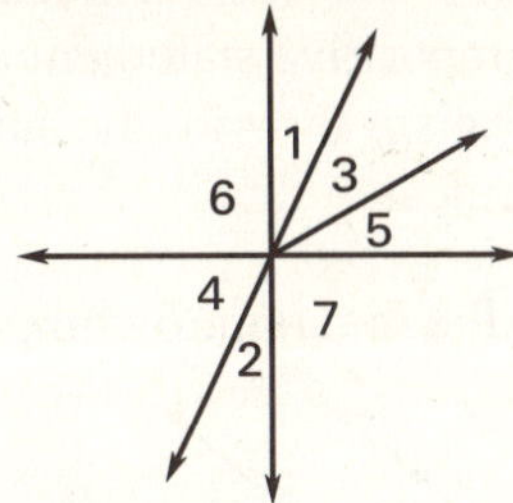

Ⓐ $\angle 3 \cong \angle 4$

Ⓑ $m\angle 7 + m\angle 2 + m\angle 4 = 180°$

Ⓒ $m\angle 3 + m\angle 5 = m\angle 4$

Ⓓ $\angle 6 \cong \angle 7$

Ⓔ $\angle 4$ and $\angle 2$ are complementary.

6. Which statement about the conditional statement "If two lines intersect, then their intersection is a point" is true?

I. The converse is true.

II. The statement can be written as a true biconditional.

III. The statement is false.

Ⓐ I only

Ⓑ I and II only

Ⓒ II and III only

Ⓓ III only

Ⓔ I, II, III

7. In the diagram, $\overline{AB} \cong \overline{CD}$. Find the length of $\overline{BD}$.

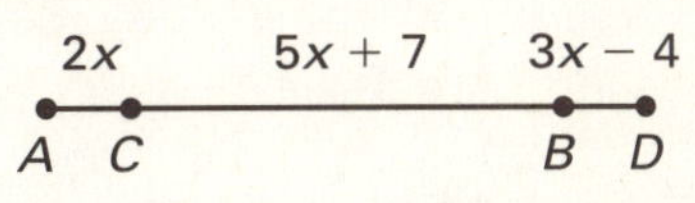

Ⓐ 8 Ⓑ 10

Ⓒ −4 Ⓓ 7

Ⓔ −7

CHAPTER 2

NAME ______________________________ DATE __________

Alternative Assessment and Math Journal

For use after Chapter 2

JOURNAL **1.** Let p be the statement "it is Saturday." Let q be the statement "you have to work at 9 A.M." Let r be the statement "you have to wake up at 7:30 A.M." Using p and q, write in *words* a conditional statement and a converse statement. Using q and r, write in symbolically a contrapositive statement and an inverse statement. Use the Law of Syllogism to write the statement that follows from the three given statements.

MULTI-STEP PROBLEM **2.** Use the diagram to complete the proof below.

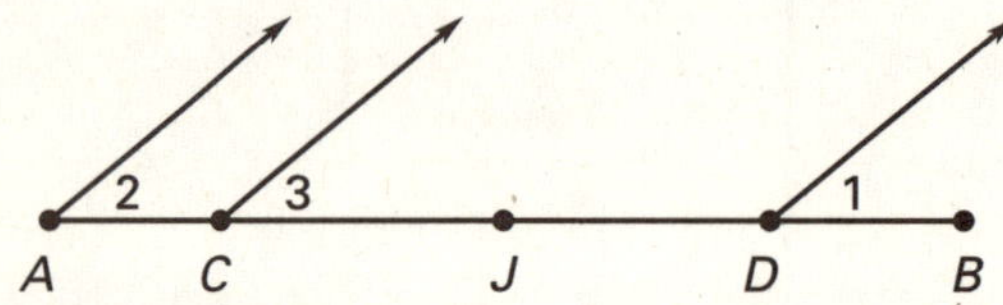

Statements	Reasons
1. $AD = BC$	**1.** ?
2. $AD = AC + CD$	**2.** ?
3. $BC = BD + CD$	**3.** ?
4. $AC + CD = BD + CD$	**4.** ?
5. $AC = BD$	**5.** ?

3. *Critical Thinking* Use the diagram from Exercise 2 to write a two column proof for each of the following.

a. Given: J is the midpoint of $\overline{AB}$.

Prove: $AJ = \frac{1}{2}AB$

b. Given: $\angle 1 \cong \angle 2$ and $\angle 2 \cong \angle 3$

Prove: $\angle 1 \cong \angle 3$

4. *Writing* Write a paragraph explaining what is a conditional statement, a converse, a contrapositive, an inverse, and a biconditional statement. Give an example of each one.

Review and Assess

Alternative Assessment Rubric

For use after Chapter 2

JOURNAL SOLUTION

1. Complete answers should include:
   - If it is Saturday, then you have to work at 9 A.M.
   - If you have to work at 9 A.M., then it is Saturday.
   - $\sim q \rightarrow \sim p$
   - $\sim p \rightarrow \sim q$
   - If it is Saturday, then you have to wake up at 7:30 A.M.

MULTI-STEP PROBLEM SOLUTION

2. Given; Segment Addition Postulate; Segment Addition Postulate; Addition property of equality; Substitution property of equality

3. a.

Statements	Reasons
1. J is the midpoint of $\overline{AB}$.	1. Given
2. $AJ = JB$	2. Definition of a midpoint
3. $AJ + JB = AB$	3. Segment Addition Postulate
4. $AJ + AJ = AB$	4. Substitution property of equality
5. $2 \cdot AJ = AB$	5. Distributive property
6. $AJ = \frac{1}{2}AB$	6. Division property of equality

b.

Statements	Reasons
1. $\angle 1 \cong \angle 2$ and $\angle 2 \cong \angle 3$	1. Given
2. $m\angle 1 \cong m\angle 2$ and $m\angle 2 \cong m\angle 3$	2. Definition of congruent angles
3. $m\angle 1 \cong m\angle 3$	3. Transitive Prop. of Congruence
4. $\angle 1 \cong \angle 3$	4. Definition of congruent angles

4. Answers may vary.

MULTI-STEP PROBLEM RUBRIC

4 Students' reasons are correct and clear. Students' proofs are correct and follow a logical order. Students' explanations are correct and clear.

3 Students' reasons are correct. Students' proofs are complete but may have one incorrect reason. Students' explanations are correct.

2 Students' reasons are complete. Students' proofs are complete but do not follow a logical order. Students' explanations are complete but not clear.

1 Students' reasons are incomplete. Students do not prove the final statement in their proofs. Students' explanations are incomplete.

NAME ______________________________ DATE ____________

Project: Investigating Advertising Slogans

For use with Chapter 2

OBJECTIVE **Explore logical reasoning as it applies to advertising slogans.**

MATERIALS newspapers, magazines, radio, or television

INVESTIGATION In this project, you will investigate the logic behind some advertising slogans. For example, what does the slogan "*Healthy Vitamins*, for a long healthy life" mean? Does it mean that if you do not take Healthy Vitamins, you will not have a long, healthy life? Is that what the advertisers would like you to think?

To analyze the logic of the slogan, first write it in *if-then* form. Also examine the inverse, converse, and contrapositive forms to help you reach conclusions about the product slogan.

Conditional: If you take Healthy Vitamins, then you will live a long, healthy life.

Inverse: If you do not take Healthy Vitamins, you will not have a long, healthy life.

Converse: If you live a long, healthy life, then you are taking Healthy Vitamins.

Contrapositive: If you do not live a long healthy life, then you did not take Healthy Vitamins.

The advertiser is hoping that you will accept the truth of their slogan and all of the conditionals associated with it.

Use a variety of media sources to investigate the advertising slogans for three different products. Analyze the logic of each slogan and identify the impact the slogan has on you or your family members.

1. Write each slogan in its original form. Identify the source it came from.
2. Write each slogan as a conditional statement in *if-then* form.
3. Write the inverse, converse, and contrapositive of each statement.
4. What do you think about each slogan after examining all of its forms? Is it a realistic slogan? Can you suggest a way to improve the slogan to make it more appealing or more realistic?
5. Tell how you or someone in your family is affected by the slogan. Does it catch your interest? Would you buy the product based only on the slogan? Explain.

PRESENT YOUR RESULTS Write a report containing the information outlined above for each slogan. Include a cut-out of the advertisement or your own drawing of the product. Describe what you learned about the use of logic in the advertising slogans you selected.

Project: Teacher's Notes

For use with Chapter 2

GOALS
- Write slogans as conditional statements in *if-then* form.
- Write and interpret the inverse, converse, and contrapositive of a conditional statement.

MANAGING THE PROJECT It is important to have a wide variety of product slogans identified. You may want to identify some product categories, such as car dealerships, beverages, or clothing, and list them on the board. Students may need help in rewriting a slogan as a conditional statement. Allow some class time for this activity, even though some work is being done outside of class. You might have students verify with you that conditional statements are written properly before they continue. Remind students that media sources for slogans should be identified as completely as possible, including names and dates of magazine sources and call letters of radio or television sources. Encourage students to discuss their slogans and ideas at home with other family members.

RUBRIC **The following rubric can be used to assess student work.**

4 The student selects and writes three advertising slogans as conditional statements. The inverses, converses, and contrapositives are written correctly. Media sources are sufficiently identified. The student thoroughly analyzes the logic, makes insightful suggestions for improvement, and describes the impact on the student or his or her family. The student writes an excellent general summary about the use of logic in advertising.

3 The student selects and writes three advertising slogans as conditional statements. Most of the inverses, converses, and contrapositives are written correctly. Media sources are for the most part sufficiently identified. For each slogan, the student may have done some part incorrectly, such as analyzing the logic, making suggestions for improvement, or describing the impact on the student or his or her family. The general summary about the use of logic in advertising is sketchy.

2 The student selects and writes at least two advertising slogans as conditional statements. The student writes the inverses, converses, and contrapositives of each conditional statement, but there are some errors. The media source of each slogan might not be identified. For each slogan, the student has done some part incorrectly. The general summary about the use of logic in advertising is poorly written.

1 The student selects and writes one or two advertising slogans as conditional statements. The student attempts to write the inverse, converse, and contrapositive of each conditional statement. The media source of each slogan is not identified correctly or may be missing.

CHAPTER 2

NAME ______________________ DATE __________

Cumulative Review

For use after Chapters 1–2

Find the distance between the two points. Round your answers to the nearest hundredth. (1.3)

1. $A(6, 2), B(0, -3)$
2. $C(-7, 1), D(4, -5)$
3. $E(-2, -2), F(5, -4)$
4. $G(0, -4), H(3, 0)$

Find the coordinates of the midpoint of a segment with the given endpoints. (1.5)

5. $A(7, 1), B(3, 3)$
6. $A(7, 1), B(-2, 6)$
7. $A(3, -7), B(-2, 1)$
8. $A(0, -4), B(3, 8)$

Find the measure of the angle without using a protractor. (1.4 and 1.6)

9. $\angle AOB$
10. $\angle DOE$
11. $\angle EOF$
12. $\angle BOF$

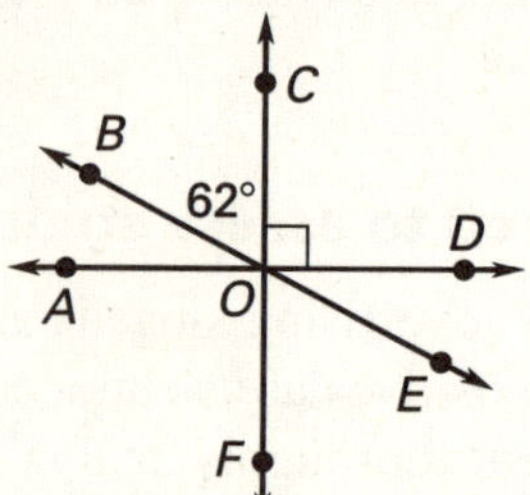

$\angle A$ and $\angle B$ are supplementary. Find $m\angle A$ and $m\angle B$. (1.6)

13. $m\angle A = 6x + 8$
    $m\angle B = 10x - 20$
14. $m\angle A = 13x - 5$
    $m\angle B = 20x + 53$
15. $m\angle A = x + 39$
    $m\angle B = 6x - 6$
16. $m\angle A = 3x + 10$
    $m\angle B = 7x + 20$

Find the perimeter (or circumference) of each figure. (Where necessary, use $\pi \approx 3.14$.) (1.7)

17.

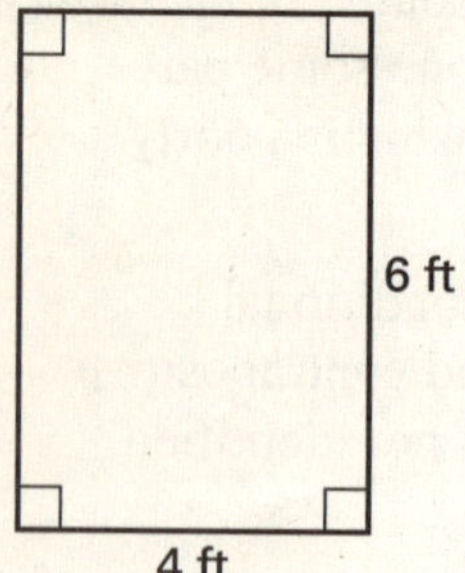

18.

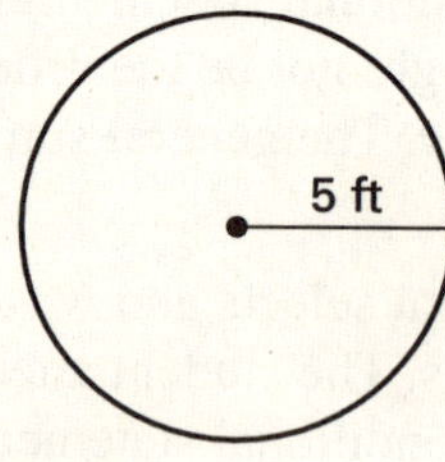

Find the inverse, converse, and contrapositive of the statement. (2.1)

19. If two angles are right angles, then they are congruent.
20. If $x = 2$, then $x^2 = 4$.

NAME ______________________________ DATE __________

Cumulative Review

For use after Chapters 1–2

Rewrite the biconditional statements as a conditional statement and its converse. (2.2)

21. Two angles are complementary if and only if their measures add to 90°.

22. Three points are collinear if and only if they lie on the same line.

Given that the statement is of form $p \rightarrow q$, write p and q. Then write the inverse and the contrapositive of $p \rightarrow q$, in words. (2.3)

23. If it is snowing, then Jennifer will ride the bus to school.

24. If the car runs out of gas, then it will not run.

Name the property used to make the conclusion. (2.4)

25. If $2x = 6$, then $x = 3$.

26. If $42° + m\angle 2 = 90°$, then $m\angle 2 = 48°$.

27. If $\overline{AB} = \overline{CD}$, then $\overline{CD} = \overline{AB}$.

28. If $m\angle D = 40°$, then $m\angle D + 50° = 90°$

29. If $m\angle J = 20°$, then $3(m\angle J) = 60°$.

30. If $\overline{WX} = \overline{YZ}$, then $\overline{WX} + \overline{AB} = \overline{YZ} + \overline{AB}$.

Use the diagram and the given information to complete the missing reasons in the proof. (2.5 and 2.6)

31. Given: lines ℓ and m

Prove: $\angle 1 \cong \angle 2$

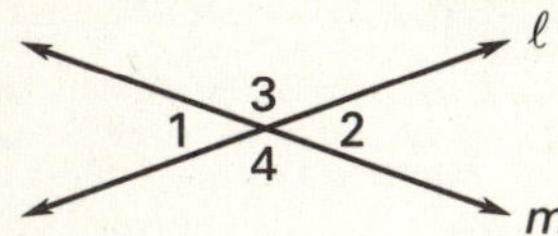

Statements	Reasons
1. lines ℓ and m	1. ____(a)____
2. $\angle 1$ and $\angle 3$ are supplementary.	2. ____(b)____
3. $\angle 2$ and $\angle 3$ are supplementary.	3. ____(c)____
4. $\angle 1 \cong \angle 2$	4. ____(d)____

32. Given: $AC = BD$

Prove: $\overline{AB} \cong \overline{CD}$

A B C D

Statements	Reasons
1. $AC = BD$	1. ____(a)____
2. $AB + BC = AC$	2. ____(b)____
3. $BC + CD = BD$	3. ____(c)____
4. $BC = BC$	4. ____(d)____
5. $AB + BC = BC + CD$	5. ____(e)____
6. $AB = CD$	6. ____(f)____
7. $\overline{AB} \cong \overline{CD}$	7. ____(g)____

ANSWERS

Chapter Support

Parent Guide

2.1: If it is December, then it is winter; false; *Sample answer*: It is summer in December in Australia. **2.2:** If two lines are perpendicular, then they form right angles; If two lines form right angles, then they are perpendicular; true; Two lines are perpendicular if and only if they form right angles. **2.3:** If Roger works, he buys new clothes. **2.4:** $m\angle K = m\angle M$

2.5: by the definition of congruent segments

2.6: $\angle ABC \cong \angle DBE$ and $\angle CBD \cong \angle ABE$

Prerequisite Skills Review

1. *Sample answer:* point X

2. *Sample answer:* point U

3. *Sample answer:* point S

4. *Sample answer:* point Z

5. 126° **6.** 126° **7.** 54° **8.** 55° **9.** 90°

10. 35° **11.** 55° **12.** 90°

13.

$m\angle 1$	36°	54°	12°	71°	68°	113°
$m\angle 2$	54°	36°	78°	19°	22°	none
$m\angle 3$	144°	126°	168°	109°	112°	67°

Strategies for Reading Mathematics

1. $\angle 2$ and $\angle 3$, $\angle 3$ and $\angle 4$, $\angle 4$ and $\angle 1$; each pair shares a common vertex and side but has no common interior points.

2. $\angle 2$ and $\angle 3$, $\angle 3$ and $\angle 4$, $\angle 4$, and $\angle 1$; the two angles in each pair are adjacent with noncommon sides that are opposite rays. **3.** $\angle 2$ and $\angle 4$; their sides form two pairs of opposite rays.

4. $\angle 5$ and $\angle 7$, $\angle 6$ and $\angle 8$; *Sample answer:* look at the relative position of the vertical angles in the drawing in the notebook and find angles that are in the same relative position.

Lesson 2.1

Warm-up Exercises

1. K **2.** K or H **3.** F or K **4.** H

Daily Homework Quiz

1. 25.1 units; 50.2 square units **2.** 20 units; 16 square units **3.** 48 units; 84 square units

4. 25 square units **5.** 75.4 m

Lesson Opener

Allow 10 minutes.

1. True; all of Massachusetts is in New England.

2. False. *Sample answer:* If you are in Connecticut, you are in New England but not in Massachusetts. **3.** True; all of Hartford is in Connecticut. **4.** False. *Sample answer:* There are other cities in Connecticut besides Hartford.

5. True; all of the beagles circle is within the dogs circle. **6.** False. *Sample answer:* Part of the dogs circle is not in the beagles circle.

7. False. *Sample answer:* Part of the animals rectangle is not in the dogs circle. **8.** True; all of the dogs circle is in the animals rectangle.

9.

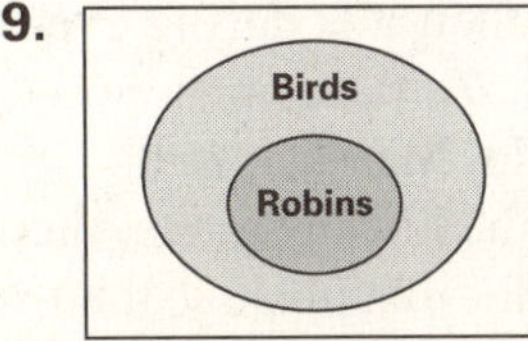

Sample answer: The diagram shows all of the robins circle inside the birds circle. There are birds that are not in the robins circle.

Practice A

1. The weather is warm; we should go swimming.

2. You want good service; take your car to Joe's Service Center. **3.** You like purple; you'll love this sweater. **4.** $2x - 12 = 40$; $x = 26$

5. The groundhog sees its shadow; there will be six more weeks of winter. **6.** If yesterday was Sunday, then today is Monday. **7.** If an object measures one foot, then it is 12 inches long.

8. If a number is divisible by 8, then it is divisible by 4. **9.** If an angle is acute, then it measures less than 90°.

Lesson 2.1 *continued*

10. If geometry is being offered, then it is during periods 2 and 4. **11.** true **12.** false; $x = \pm 4$ **13.** true **14.** true **15.** false;

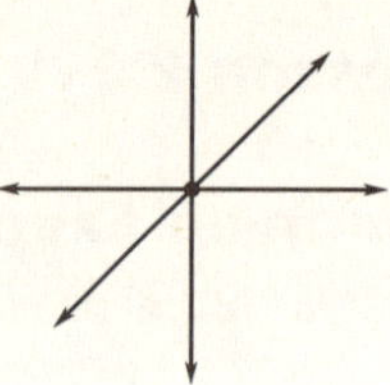

16. If you play on the tennis team, then you like tennis; If you don't play on the tennis team, then you don't like tennis. **17.** If $2x$ is even, then x is odd; If $2x$ is not even, then x is not odd. **18.** If $\angle P$ is acute, then $m\angle P = 45°$; If $\angle P$ is not acute, then $m\angle P \neq 45°$. **19.** Through any three noncollinear points there exists exactly one plane. **20.** Through any two points there exists exactly one line. **21.** If two planes intersect, then their intersection is a line. **22.** If two points lie in a plane, then the line containing them lies in the plane.

Practice B

1. If it is 6 P.M., then it is time for dinner.

2. If the carton is full, then there are 12 eggs.

3. If a number is divisible by 2 and 3, then it is divisible by 6. **4.** If an angle is obtuse, then it measures more than 90° and less than 180°.

5. If Geometry is offered, then it is during even numbered blocks. **6.** true **7.** false; $x = \pm 6$ **8.** true **9.** false; Springfield, Massachusetts **10.** true **11.** If you go to the hockey game, then you like hockey; If you do not like hockey, then you do not go to the hockey game; If you do not go to the hockey game, then you do not like hockey. **12.** If $3x$ is odd, then x is odd. If x is not odd, then $3x$ is not odd; If $3x$ is not odd, then x is not odd. **13.** If $\angle P$ is a right angle, then $m\angle P = 90°$; If $m\angle P \neq 90°$, then $\angle P$ is not a right angle; If $\angle P$ is not a right angle, then $m\angle P \neq 90°$. **14.**

ℓ
A m

15.

A B
C

16.

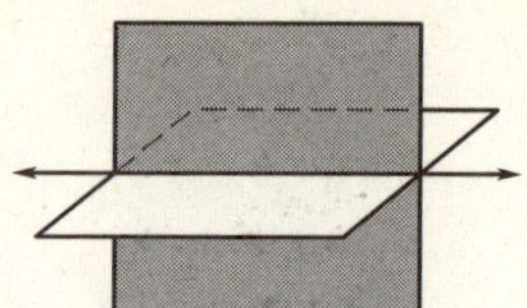

17. Through any three noncollinear points there exists exactly one plane. **18.** Through any two points there exists exactly one line. **19.** If two planes intersect, then their intersection is a line.

20. If two points lie in a plane, then the line containing them lies in the plane.

Practice C

1. If I get all of my homework done, then I will go to the game. **2.** If the temperature is 10°F, then the water will freeze. **3.** If a student is on the high honor roll, then they have at least a 90 average. **4.** If Bert goes shopping for groceries, then it is Wednesday. **5.** If a number is even, then 2 is a factor of it. **6.** true **7.** false; $\left(\frac{1}{2}\right)^2 < \frac{1}{2}$ **8.** false; $|4 + -2| \neq |4| + |-2|$

9. true **10.** false;

11. If you like to be at the beach, then you like volleyball. (False); If you do not like volleyball, then you do not like to be at the beach. (False); If you do not like to be at the beach, then you do not like volleyball. (False) **12.** If x is odd, then $4x$ is even. (True); If $4x$ is not even, then x is not odd. (False); If x is not odd, then $4x$ is not even. (False) **13.** If $\angle P$ is obtuse, then $m\angle P = 109°$. (False); If $m\angle P \neq 109°$, then $\angle P$ is not obtuse. (False); If $\angle P$ is not obtuse, then $m\angle P \neq 109°$. (True)

14.

A B

15.

A B
C

16.

D
E
F

17. If two lines intersect, then their intersection is exactly one point. **18.** Through any three noncollinear points there exists exactly one plane.

19. If two planes intersect, then their intersection is a line. **20.** Through any two points there exists exactly one line.

Lesson 2.1 *continued*

Reteaching with Additional Practice

1. a. If you will wash the dishes, then I will dry them. **b.** If you will not wash the dishes, then I will not dry them. **c.** If I will dry the dishes, then you will wash them. **d.** If I will not dry the dishes, then you will not wash them.

2. a. If a square has side length of 3 cm, then it has an area of 9 square cm. **b.** If a square does not have side length of 3 cm, then it does not have an area of 9 square cm. **c.** If a square has an area of 9 square cm, then it has side length of 3 cm. **d.** If a square does not have an area of 9 square cm, then it does not have a side length of 3 cm. **3. a.** If an angle has a measure of 90°, then it is a right angle. **b.** If an angle does not have a measure of 90°, then it is not a right angle. **c.** If an angle is a right angle, then it has a measure of 90°. **d.** If an angle is not a right angle, then it does not have a measure of 90°.

4. true **5.** False. In the diagram shown, $\overleftrightarrow{AB}$ and C are in the same plane.

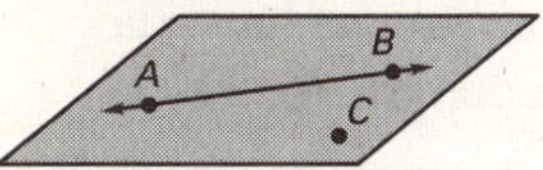

Interdisciplinary Application

1. 3 **2.** none **3.** 3 **4.** none

Challenge: Skills and Applications

1. inverse: If a bicycle is not red, then it does not have two seats. converse: If a bicycle has two seats, then it is red. contrapositive: If a bicycle does not have two seats, then it is not red.

2. inverse: If a thopfoo cannot make a pucho, then all gorups are not purple. converse: If all gorups are purple, then a thopfoo can make a pucho. contrapositive: If all gorups are not purple, then a thopfoo cannot make a pucho.

3. inverse: If a student does not enjoy mathematics, then he or she does not have a logical mind. converse: If a student has a logical mind, then he or she enjoys mathematics. contrapositive: If a student does not have a logical mind, then he or she does not enjoy mathematics.

4. inverse: If a person is not a citizen of the Regal Kingdom, then he or she does not have blue hair. converse: If a person has blue hair, then he or she is a citizen of the Regal Kingdom. contrapositive: If a person does not have blue hair, then he or she is not a citizen of the Regal Kingdom.

5. true; If a positive integer is divisible by both 3 and 5, then it is divisible by 15; true.

6. false; If a positive integer is divisible by 12, then it is divisible by both 2 and 3; true; counterexample for original statement: 18 is divisible by 2 and 3 but not by 12. **7.** true; If an integer n is divisible by 3, then n^2 is divisible by 3; true.

8. false; If an integer n is divisible by 4, then n^2 is divisible by 4; true; counterexample for original statement: If $n = 6$ then n^2 is divisible by 4 but n is not divisible by 4. **9.** true; Postulate 11: If two planes intersect, then their intersection is a line

10. false; There are many planes (but only one line) through any pair of points. **11.** true; Since a plane contains at least 3 *noncollinear* points, it certainly contains at least three points.

12. false; Any three collinear points are also coplaner. **13.** *Sample answer:* There exists exactly one plane which contains both lines a and b. **14.** Postulate 6: A line contains at least two points. **15.** Postulate 7: If two lines intersect, then their intersection is exactly one point.

16. Postulate 8: Through any three noncollinear points there exists exactly one plane.

17. Postulate 10: If two points lie in a plane, then the line containing them lies in the plane.

18. Answers will vary.

Lesson 2.2

Warm-up Exercises

1. If $AM = MB$, then M is the midpoint of $\overline{AB}$.

2. If points A, B, and C are collinear, then $AB + BC = AC$. **3.** If $x = 6$ or -6, then $|x| = 6$. **4.** If the ball crosses the goal line, then you score a goal.

Daily Homework Quiz

1. If a snake is a constrictor, then it is not poisonous. **2.** If a snake is not poisonous, then it is a constrictor. **3.** If a snake is poisonous, then it is not a constrictor. **4.** hypothesis: a snake is a constrictor; conclusion: it is not poisonous

Lesson 2.2 *continued*

Lesson Opener

Allow 15 minutes.

1. Yes. *Sample answer:* The sum of the angle measures of a triangle are 180°, and 180° − 90° = 90°. Therefore, the two angles that are not 90° must add up to 90°. **2.** All of the triangles have one right angle and two complementary angles. **3.** If a triangle has two complementary angles, then it has one right angle; yes; the sum of two complementary angles is 90°, so the third angle must have a measure of 90° in order for the sum of the angle measures to be 180°. **4.** All of the triangles have two complementary angles and one right angle.

5. Yes. *Sample answer:* A polygon with four angles is a quadrilateral and a quadrilateral with four right angles is a rectangle. **6.** All of the polygons have four right angles and they are all rectangles. **7.** If a polygon is a rectangle, then it has four right angles; yes; by definition of a rectangle. **8.** All of the polygons are rectangles and they all have four right angles.

Technology Activity

1. Conditional: if the three sides of a triangle are congruent, then the three angles are congruent; Converse: if the three angles of a triangle are congruent, then the three sides are congruent.

2. Answers will vary.

Practice A

1. false **2.** true **3.** true **4.** false **5.** true

6. false **7.** false **8.** If two segments are congruent, then they have the same measure; If two segments have the same measure, then they are congruent. **9.** If three points are collinear, then they lie on the same line; If three points lie on the same line, then they are collinear. **10.** If four points lie in the same plane, then they are coplanar; If four points are coplanar, then they lie in the same plane. **11.** If you want to go to the movies Friday night, then you have to clean your room; If you clean your room, then you may go to the movies Friday night. **12.** If you want to become president of the United States, then you must be 35 years old; If you are 35 years old, then you may become president of the United States. **13.** Lansing, MI **14.** 20°

15. Dalmatian **16.** June

17.

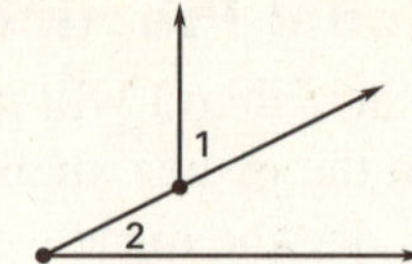

18. A saxophone that has a frequency of 104 cycles per second to 622 cycles per second is called a B-flat tenor saxophone.

19. A saxophone that has a frequency of 138 cycles per second to 831 cycles per second is called a E-flat alto saxophone.

Practice B

1. true **2.** true **3.** true **4.** false **5.** true

6. false **7.** false **8.** If $x = 4$, then $x^2 = 16$; If $x^2 = 16$, then $x = 4$. **9.** If point Y lies between points X and Z, then $XY + YZ = \overline{XZ}$; If $XY + YZ = XZ$, then point Y lies between points X and Z. **10.** If the car runs, then there is gas in the tank; If there is gas in the tank, then the car runs. **11.** If two angles are congruent, then they have the same measure; If two angles have the same measure, then they are congruent.

12. If an angle is a right angle, then it measures 90°; If an angle measures 90°, then it is a right angle. **13.** If you live in Michigan, then you live in Detroit. **14.** If an angle is acute then it measures 30°. **15.** If the sum of two angles is 180°, then they are supplementary; Two angles are supplementary if and only if their sum is 180°. **16.** If two angles have the same measure, then they are congruent; Two angles are congruent if and only if they have the same measure.

17. If two angles are not adjacent, then they are vertical. **18.** A saxophone that has a frequency of 69 cycles per second to 416 cycles per second is called an E-flat baritone saxophone.

19. A saxophone that has a frequency of 104 cycles per second to 622 cycles per second is called a B-flat tenor saxophone.

20. A saxophone that has a frequency of 138 cycles per second to 831 cycles per second is called a E-flat alto saxophone.

Practice C

1. false **2.** true **3.** true **4.** true **5.** true

Lesson 2.2 *continued*

6. false **7.** false **8.** If an angle is acute, then it measures less than 90°; If an angle measures less than 90°, then it is acute. **9.** If three points are collinear, then they lie on the same line; If three points lie on the same line, then they are collinear. **10.** If it is Friday night, then I eat pizza; If I eat pizza, then it is Friday night.

11. If it rains, then the game is cancelled; If the game is cancelled, then it is raining. **12.** If a number is divisible by 6, then it is divisible by 2 and 3; If a number is divisible by 2 and 3, then it is divisible by 6. **13.** If you live in Michigan, then you live in Detroit; Lansing, MI **14.** If an angle is acute, then it measures 30°; 45°

15. If the sum of two angles is 180°, then they are supplementary; Two angles are supplementary if and only if their sum is 180°. **16.** If two angles have the same measure, then they are congruent; Two angles are congruent if and only if they have the same measure.

17. If two angles are not adjacent, then they are vertical;

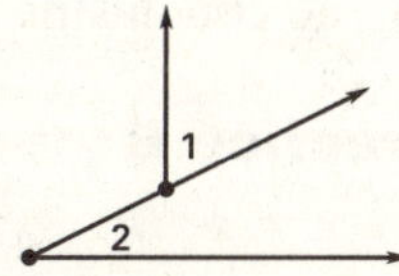

18. A saxophone that has a frequency of 69 cycles per second to 416 cycles per second is called an E-flat baritone saxophone.

19. A saxophone that has a frequency of 104 cycles per second to 622 cycles per second is called a B-flat tenor saxophone.

20. A saxophone that has a frequency of 138 cycles per second to 831 cycles per second is called an E-flat alto saxophone. **21.** It is an E-flat baritone saxophone. **22.** nothing; It could be any of the three saxophones.

Reteaching with Additional Practice

1. conditional statement: If a number is a perfect square, then it is the product of some number times itself; converse: If a number is the product of some number and itself, then it is a perfect square. **2.** conditional statement: If two angles are complementary, then their sum is 90°; converse: If two angles have a sum of 90°, then they are complementary. **3.** conditional statement: If a real number is rational, then it can be written in the form $\frac{p}{q}$, where p and q are integers and $q \neq 0$; converse: If a real number can be written in the form $\frac{p}{q}$, where p and q are integers and $q \neq 0$, then it is a rational number.

4. a. The statement is a biconditional statement because it contains the words "if and only if." **b.** false **5. a.** The statement is a biconditional statement because it contains the words "if and only if." **b.** true **6.** If angles are vertical angles, then the sides of the two angles form two pairs of opposite rays; True; The sides of two angles form two pairs of opposite rays if and only if the angles are vertical angles. **7.** If a is 0 or b is 0, then the product ab is 0; True; The product of ab is 0 if and only if either a is 0 or b is 0.

Real-Life Application

1. If you were assigned an office detention, then you were late to some class. **2.** If you were assigned a Saturday detention, then you were insubordinate to an adult. **3.** If you lost your computer privileges for one week, you were using the internet without authorized access. **4.** If you and a classmate are required to attend conflict resolution sessions, then you two were involved in a verbal argument. **5.** If you served 3 days of ISS, then you skipped a day of school.

In 1–5, the converses are not always true. Therefore none of the conditional statements and the converses listed are biconditional statements.

Challenge: Skills and Applications

1.

p	q	$q \to p$
T	T	T
T	F	T
F	T	F
F	F	T

2.

p	q	$\sim p$	$\sim q$	$\sim p \to \sim q$
T	T	F	F	T
T	F	F	T	T
F	T	T	F	F
F	F	T	T	T

Lesson 2.2 *continued*

3.

p	q	$\sim q$	$\sim p$	$\sim q \to \sim p$
T	T	F	F	T
T	F	T	F	F
F	T	F	T	T
F	F	T	T	T

4.

p	q	$p \leftrightarrow q$
T	T	T
T	F	F
F	T	F
F	F	T

5. a.

p	q	$\sim p$	$\sim q$	$\sim p \leftrightarrow \sim q$
T	T	F	F	T
T	F	F	T	F
F	T	T	F	F
F	F	T	T	T

b. $p \leftrightarrow q$

Lesson 2.3

Warm-up Exercises

1. If the golf tournament is not canceled, it does not thunder. **2.** If x does not equal 0 or 1, then $x(x - 1) \neq 0$. **3.** If $\angle A$ is not congruent to $\angle B$, then $m\angle A \neq 45°$.

Daily Homework Quiz

1. If we go to the beach, then it is sunny; If it is sunny, then we will go to the beach.

2. A rhombus has four equal sides but is not a square. **3.** A bicycle has wheels but is not a car.

4. No; x can be any number greater than 4.

Lesson Opener

Allow 15 minutes.

1. Check work.

2. A → B; false

B → A; false

A → not B; true

B → not A; true

not A → B; true

not B → A; true

not A → not B; false

not B → not A; false

3. Yes; not A and B, A and not B

4. not A ↔ B, true; A ↔ not B, true; *Sample answer:* A biconditional is true when both the conditional and its converse are true.

Practice A

1. The sky is not cloudy. **2.** It is not raining.

3. If it is raining, then the sky is cloudy.

4. If it is not raining, then the sky is not cloudy.

5. If the sky is cloudy, then it is raining.

6. If the sky is not cloudy, then it is not raining.

7. Law of Detachment **8.** invalid

9. Law of Detachment **10.** Law of Syllogism

11. invalid **12.** Law of Syllogism **13.** Tim will go to the hospital. **14.** Hank will buy a new car. **15.** Plane *A* and plane *B* intersect in a line.

16. You received a detention. **17.** no conclusion

18. no conclusion

Practice B

1. $x \neq 4$ **2.** $3x + 2 \neq 14$ **3.** If $3x + 2 = 14$, then $x = 4$. **4.** If $3x + 2 \neq 14$, then $x \neq 4$.

5. If $x = 4$, then $3x + 2 = 14$. **6.** If $x \neq 4$, then $3x + 2 \neq 14$. **7.** Law of Detachment

8. invalid **9.** Law of Detachment

10. Law of Syllogism **11.** invalid **12.** Law of Syllogism **13.** If the dog did not lick Susan, then Carl did not tell the dog to "kiss." **14.** If the dog does not lick Susan, then Susan will not scream. **15.** If the dog runs away, then Susan will scream. **16.** If Carl tells the dog to "kiss," then the dog will lick Susan. If the dog licks Susan, then Susan will scream. If Susan screams, then the dog will run away. **17.** yes; Carl tells the dog to "kiss." So the dog licks Susan.

18. yes; Carl tells the dog to "kiss." The dog licks Susan. Susan screams. The dog runs away.

Practice C

1. $x \neq -4$; false **2.** $3x + 2 \neq -10$; false

3. If $3x + 2 = -10$, then $x = -4$; true

4. If $3x + 2 \neq -10$ then $x \neq -4$; true

5. If $x = -4$, then $3x + 2 = -10$; true

Lesson 2.3 *continued*

6. If $x \neq -4$, then $3x + 2 \neq -10$; true

7. Law of Syllogism **8.** Law of Detachment

9. invalid **10.** invalid **11.** Law of Detachment **12.** Law of Syllogism **13.** If it is raining, then the apartment ceiling is leaking. **14.** If it is Tuesday, then it is raining. **15.** If it is Tuesday, then it is raining. $(t \rightarrow s)$ If it is raining, then the apartment ceiling is leaking. $(s \rightarrow r)$ If the apartment ceiling leaks, then I will call the supervisor. $(r \rightarrow p)$ If I call the supervisor, then I must pay the rent. $(p \rightarrow q)$ **16.** yes; It is Tuesday, so it is raining. The apartment ceiling leaks. I call the supervisor. I pay the rent.

17. no; It may rain on a day other than Tuesday.

Reteaching with Additional Practice

1. a. If two lines are parallel, then they do not intersect. **b.** If two lines do not intersect, then they are parallel. **c.** true **2. a.** If you are in North America, then you are in the United States.

b. If you are in the United States, then you are in North America. **c.** false **3. a.** If two angles are not supplementary, then the measures do not sum to 180°. **b.** If two angles do not sum to 180°, then they are not supplementary.

4. a. If a number is not divisible by 5, then it is not divisible by 10. **b.** If a number is not divisible by 10, then it is not divisible by 5. **5.** False. We do not know whether or not Pete is thirsty.

Interdisciplinary Application

1. If the batter does not become a base runner, then the batter makes an out. **2.** If the batter becomes a base runner, then the batter does not make an out. **3.** If the batter does not make an out, then the batter becomes a base runner.

4. If the score is tied at the end of nine innings, then play continues into extra innings.

5. If the play continues into extra innings, then the score is tied at the end of nine innings.

6. If the score is not tied at the end of nine innings, then play does not continue into extra innings. **7.** Your team wins the game.

8. (1) If a batter crosses home plate, then the game will be tied. (2) If the outfielder makes the catch, then the inning is over.

Math and History Application

1. 12 hours, or 720 minutes **2.** 720 days

3. $\approx$ 2 years **4.** twice a day; Clock B

Challenge: Skills and Applications

1. a. Bradley's hat and Atsuko's hat are not both blue. **b.** Atusko's hat is yellow. *Sample answer:* If her hat were blue, Bradley would have known that his hat is yellow. **c.** no **2.** Linda Ricardo and Kevin Ricardo live in Chicago; Mary Tanaka and Ivan Tanaka live in Boston; Nancy Smith and Jeff Smith live in Denver. **3. a.** 1 **b.** If the coins that are turned are both heads or both tails, turning them over changes the number of heads by 2. If the coins that are turned do not match, turning them over does not change the number of heads. So, the number of heads remains an odd number. **c.** Each successive turn produces an odd number of heads.

Quiz 1

1. The statement is already in if-then form; converse: If next month is October, then this month is September. Both the statement and its converse are true, so they can be combined to form a biconditional statement: This month is September if and only if next month is October.

2. if-then form: If a time period is a decade, then it is a period of ten years; converse: If a time period is ten years, then it is a decade. The statement and the converse cannot form a biconditional statement because the converse is not necessarily true. **3.** if-then form: If I am sleeping, then I am breathing; converse: If I am breathing, then I am sleeping. The statement and its converse cannot form a biconditional statement because the converse is not necessarily true.

4. Yes; your team scored four touchdowns, therefore your team won. **5.** Yes; your team scored 4 touchdowns, therefore your team won, and therefore, the coach is happy.

Lesson 2.4

Warm-up Exercises

1. 9 **2.** -23 **3.** 27 **4.** 9 **5.** -4

Lesson 2.4 *continued*

Daily Homework Quiz

1. This animal has no stripes. **2.** $\sim p \rightarrow \sim q$; if this animal does not have stripes, then it is not a tiger. **3.** $\sim q \rightarrow \sim p$; if this animal is not a tiger, then it does not have stripes. **4.** If it is Friday, then tomorrow we will swim.

Lesson Opener

Allow 10 minutes.

1.

Statement	Reason
$10y + 5 = 25$	Given
$10y + 5 - 5 = 25 - 5$	Subtract 5 from both sides.
$10y = 20$	Simplify.
$\frac{10y}{10} = \frac{20}{10}$	Divide both sides by 10.
$y = 2$	Simplify.

2.

Statement	Reason
$6x + 3 = 9(x - 1)$	Given
$6x + 3 = 9x - 9$	Distribute 9.
$6x - 9x + 3 = 9x - 9x - 9$	Subtract $9x$ from both sides.
$-3x + 3 = -9$	Simplify.
$-3x + 3 - 3 = -9 - 3$	Subtract 3 from both sides.
$-3x = -12$	Simplify.
$\frac{-3x}{-3} = \frac{-12}{-3}$	Divide both sides by -3.
$x = 4$	Simplify.

Practice A

1. D **2.** A **3.** G **4.** B **5.** C **6.** F **7.** E **8.** 15 **9.** $\frac{1}{2}$ **10.** AF **11.** $m\angle MJC = m\angle DCF$ **12.** DB, $YZ = JK$ **13.** 5(3) **14. a.** Distributive property **b.** Subtraction property of equality **c.** Addition property of equality **15. a.** Subtraction property of equality **b.** Subtraction property of equality **c.** Division property of equality

16. a. Reflexive property of equality **b.** Subtraction property of equality

17.

Statement	Reason
$5(2x - 1) = 9x + 4$	Given
$10x - 5 = 9x + 4$	Distributive property
$x - 5 = 4$	Subtraction property of equality
$x = 9$	Addition property of equality

18.

Statement	Reason
$-4(x - 5) = 13$	Given
$-4x + 20 = 13$	Distributive property
$-4x = -7$	Subtraction property of equality
$x = \frac{7}{4}$	Division property of equality

Practice B

1. $m\angle T$ **2.** MN, $KL = RW$ **3.** 22 **4.** $RL = BC$ **5.** 3(45) **6.** $\frac{1}{3}$ **7. a.** Distributive property **b.** Addition property of equality **c.** Subtraction property of equality **8. a.** Addition property of equality **b.** Addition property of equality **c.** Division property of equality **9. a.** Segment Addition Postulate **b.** Substitution property of equality **c.** Distributive property **10. a.** Reflexive property of equality **b.** Addition property of equality **c.** Angle Addition Postulate **d.** Angle Addition Postulate **e.** Substitution property of equality **11. a.** Definition of perpendicular segments and definition of right angle **b.** Definition of perpendicular segments and definition of right angle **c.** Substitution or transitive property of equality

Practice C

1. $\frac{1}{3}$ **2.** $m\angle DEF$, $m\angle ABC = m\angle STO$ **3.** 59° **4.** HS **5.** 5(7.5) **6.** $m\angle DBA = m\angle MCS$ **7. a.** Distributive property **b.** Subtraction property of equality **c.** Subtraction property of equality **d.** Division property of equality **8. a.** Distributive property **b.** Simplify. **c.** Subtraction property of equality **d.** Symmetric property **9. a.** Reflexive property of equality **b.** Addition property of equality **c.** Segment Addition Postulate **d.** Segment Addition Postulate **e.** Substitution property of equality

Lesson 2.4 *continued*

10. $\overleftrightarrow{AB} \perp \overleftrightarrow{EF}$; $\overleftrightarrow{CD} \perp \overleftrightarrow{EF}$ — Given

$\angle 1$ and $\angle 2$ are right angles — Def. of $\perp$ lines

$m\angle 1 = 90°$; $m\angle 2 = 90°$ — Def. of right angles

$m\angle 1 = m\angle 2$ — Substitution property of equality

11. $m\angle 1 = 40°, m\angle 2 = 50°$ — Given

$m\angle 1 + m\angle 2 = 90°$ — Addition property of equality

$\angle 1$ and $\angle 2$ are complementary. — Def. of complementary angles

Reteaching with Additional Practice

1.

$3 = 5x$	Subtraction prop. of equality
$\frac{3}{5} = x$	Division prop. of equality

2.

$4 + 6x + 10 = 11 - x$	Distributive prop.
$6x + 14 - 11 - x$	Simplify.
$6x = -3 - x$	Subtr. prop. of equality
$7x = -3$	Add. prop. of equality
$x = -\frac{3}{7}$	Division prop. of equality

3.

$6x - 2 = -4x + 4$	Distributive prop.
$6x = -4x + 6$	Addition prop. of equality
$10x = 6$	Addition prop. of equality
$x = \frac{3}{5}$	Division prop. of equality

4.

$\frac{1}{5}x + \frac{17}{5} = 2x$	Subtraction prop. of equality
$\frac{17}{5} = \frac{9}{5}x$	Subtraction prop. of equality
$\frac{17}{9} = x$	Multiplication prop. of equality

5.

$MN = PQ$	Given
$MP = MN + NP$	Segment Addition Postulate
$MP = PQ + NP$	Substitution prop. of equality
$NQ = PQ + NP$	Segment Addition Postulate
$MP = NQ$	Substitution prop. of equality

6.

$AB = DE$	Given
$AD = AB + BD$	Segment Addition Postulate
$AD = DE + BD$	Substitution prop. of equality
$BE = DE + BD$	Segment Addition Postulate
$AD = BE$	Substituation prop. of equality

7.

$m\angle AQB = m\angle CQD$	Given
$m\angle AQC = m\angle AQB + m\angle BQC$	Angle Addition Postulate
$m\angle AQC = m\angle CQD + m\angle BQC$	Substitution prop. of equality
$m\angle BQD = m\angle CQD + m\angle BQC$	Angle Addition Postulate
$m\angle AQC = m\angle BQD$	Substitution prop. of equality

8.

$m\angle RPS = m\angle TPV$	Given
$m\angle TPV = m\angle SPT$	Given
$m\angle RPS = m\angle SPT$	Transitive prop. of equality
$m\angle RPV = m\angle RPS + m\angle SPT + m\angle TPV$	Angle Addtion Postulate
$m\angle RPV = m\angle RPS + m\angle RPS + m\angle RPS$	Substitution prop. of equality
$m\angle RPV = 3(m\angle RPS)$	Simplify.

Cooperative Learning Activity

1.

1, 2, 1, 3, 2	Given
2, 1, 1, 3, 2	Rule 3
2, 3, 2	Rule 2
3, 2, 2	Rule 3
3, 2	Rule 1

2.

1, 2, 2, 2, 2, 1	Given
1, 3, 1	Rule 2
3, 1, 1	Rule 3
3, 2	Rule 1

Lesson 2.4 *continued*

3. 1, 2, 3, 4, 4, 2, 3 Given
1, 3, 2, 4, 4, 2, 3 Rule 3
3, 1, 2, 4, 4, 2, 3 Rule 3
3, 3, 3 Rule 2
3, 2 Rule 1

4. 1, 4, 4, 3, 4 Given
4, 1, 4, 3, 4 Rule 3
4, 3, 4 Rule 2
3, 4, 4 Rule 3
3, 2 Rule 1

5. 1, 4, 3, 2, 2, 3, 4, 2, 1, 4, 4, 4 Given
4, 1, 3, 2, 2, 3, 4, 2, 1, 4, 4, 4 Rule 3
4, 3, 4 Rule 2
3, 4, 4 Rule 3
3, 2 Rule 1

Real-Life Application

1. a. 68°F **b.** 86°F **c.** 23°F

2.

$F = \frac{9}{5}C + 32$	Given
$F - 32 = \frac{9}{5}C$	Subtraction property of equality
$\frac{5}{9}(F - 32) = C$	Multiplication prop. of equality

3. $\frac{5}{9}F - \frac{160}{9} = C$; Distributive property

4. a. 35°C **b.** −10°C **c.** −27.5°C

Challenge: Skills and Applications

1. a. $4 \cdot 6$ is 1 less than 5^2. **b.** $5 \cdot 7$ is 1 less than 6^2. **c.** $n(n + 2)$ is 1 less than $(n + 1)^2$.

d. *Sample answer:* $n(n + 2) = n^2 + 2n$
$= (n^2 + 2n + 1) - 1 = (n + 1)^2 - 1$

2. a. $\frac{1}{2}, \frac{2}{3}, \frac{3}{4}$ **b.** $\frac{n}{n+1}$

c. *Sample answer:* By substituting $\frac{1}{n} - \frac{1}{n+1}$ for $\frac{1}{n(n+1)}$ and applying the associative property of addition, the sum telescopes to $\frac{n}{n+1}$ as follows: $\frac{1}{1 \cdot 2} + \frac{1}{2 \cdot 3} + \frac{1}{3 \cdot 4} + \ldots + \frac{1}{n(n+1)}$

$= \left(1 - \frac{1}{2}\right) + \left(\frac{1}{2} - \frac{1}{3}\right) + \left(\frac{1}{3} - \frac{1}{4}\right) + \ldots + \left(\frac{1}{n} - \frac{1}{n+1}\right)$

$= 1 + \left(-\frac{1}{2} + \frac{1}{2}\right) + \left(-\frac{1}{3} + \frac{1}{3}\right) + \ldots + \left(-\frac{1}{n} + \frac{1}{n}\right) - \frac{1}{n+1}$

$= 1 - \frac{1}{n+1}$

$= \frac{n+1}{n+1} - \frac{1}{n+1}$

$= \frac{n}{n+1}$

3. a. *Sample answer:* Write the two integers as n and $n + 1$. The sum is $2n + 1$, which is an odd number because it is 1 more than the even number $2n$. **b.** *Sample answer*: Write the three integers as n, $n + 1$, and $n + 2$. The sum is $3n + 3$, or $3(n + 1)$, which is a multiple of 3.

4. a. *Sample answer:* $(2n)^2 = 4n^2 = 2(2n^2)$, which is an even integer. **b.** *Sample answer:* $(2n + 1)^2 = 4n^2 + 4n + 1 = 2(2n^2 + 2n) + 1$, which is an odd integer.

5. reflexive, transitive **6.** symmetric

7. reflexive, symmetric, transitive **8.** transitive

Lesson 2.5

Warm-up Exercises

1. Symmetric prop. of equality **2.** Reflexive prop. of equality **3.** Transitive prop. of equality

4. Substitution prop. of equality

Daily Homework Quiz

1. *ay* 2. $HJ = FG$ 3. $BC = YZ$

4.

Statements	Reasons
1. $3m - 2 = m + 8$	**1.** Given
2. $2m - 2 = 8$	**2.** Subtraction prop. of equality
3. $2m = 10$	**3.** Addition prop. of equality
4. $m = 5$	**4.** Division prop. of equality

Lesson Opener

Allow 10 minutes.

Statements

2. 81 is between 80 and 89.

Reasoning

3. Definition of a B **4.** 80 and 81 are the least two percents in the interval 80–89.

5. Definition of $-$.

Technology Activity

1. Check student's work.

2. Check student's work.

3. Check student's work.

Practice A

1. C **2.** B **3.** A **4.**

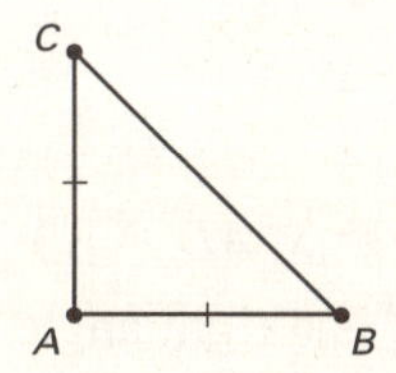

5. **6.**

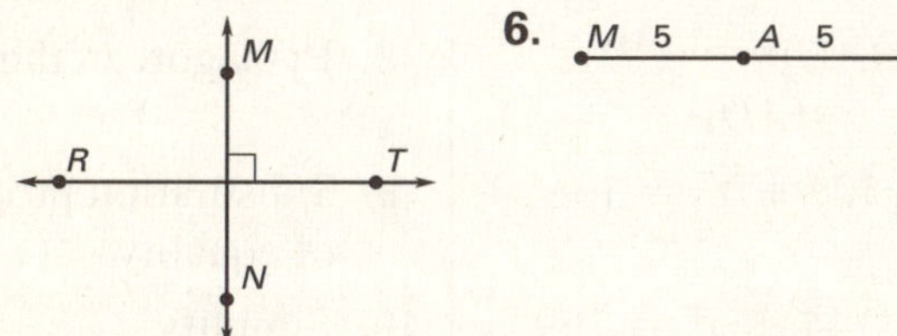

7.

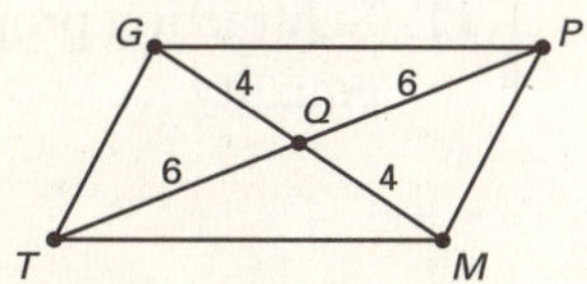

8. S A R

9. B N D M

10. 1. Given 2. Given 3. Transitive Property of Segment Congruence **11.** 2. Segment Addition Postulate 3. Segment Addition Postulate
4. Substitution Property of Equality

12. You are given $\overline{AD}$. The Segment Addition Postulate says that $AD = AB + BD$ and also that $BD = BC + CD$. You can substitute the last equation into the first and conclude that $AD = AB + BC + CD$.

Practice B

1. C **2.** A **3.** B

4.

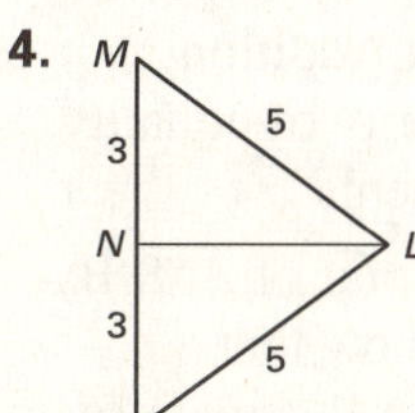

5.

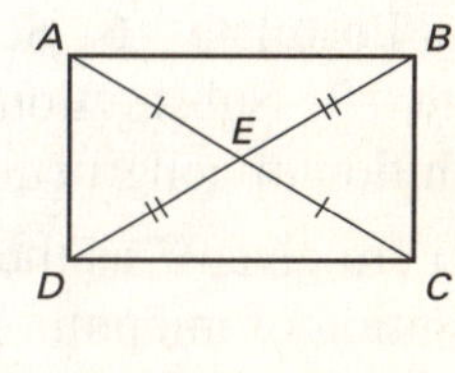

6.

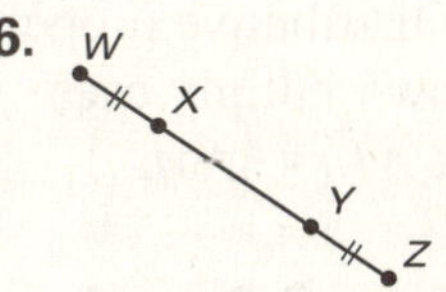

7. 1. Given 2. Given 3. Substitution property equality 4. Definition of congruent segments 5. Given 6. Transitive Property of Congruence

8. 1. Given 2. Reflexive property of equality 3. Addition property of equality 4. Segment Addition Postulate 5. Segment Addition Postulate 6. Substitution property of equality

9. You are given that $\overline{HI}$ and $\overline{IJ}$ each equal 8 so by substitution they are equal to each other. If they are equal, by definition $\overline{HI} \cong \overline{IJ}$. Since $\overline{IJ}$ is also congruent to $\overline{JH}$, given, we conclude $\overline{HI} \cong \overline{JH}$ by the Transitive Property of Congruence.

Practice C

1.

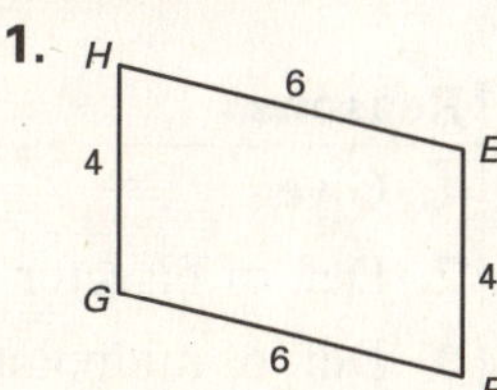

2.

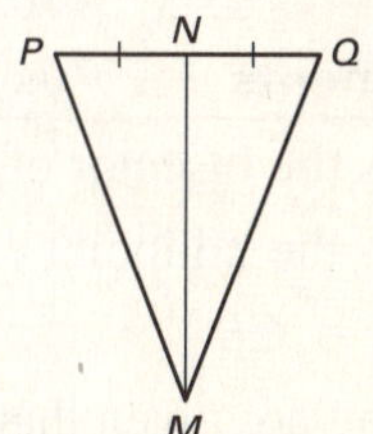

Lesson 2.5 *continued*

3. 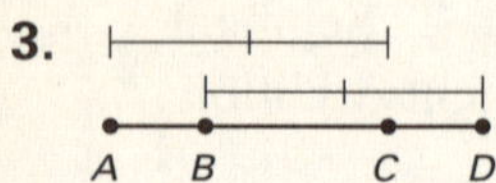

4. 1. Given 2. Transitive property of equality 3. Definition of congruent segments

5. 1. Given 2. Substitution property of equality 3. Definition of congruent segments 4. Given 5. Transitive Property of Congruence

6. 1. Given 2. Definition of congruent segments 3. Reflexive property of equality 4. Addition property of equality 5. Segment Addition Postulate 6. Segment Addition Postulate 7. Substitution property of equality 8. Definition of congruent segments

7. You are given T as the midpoint of $\overline{AM}$. By the definition of midpoint you know that $AT = TM$. The Segment Addition Postulate says $AT + TM = AM$. Since $AT = TM$, substitute and get $AT + AT = AM$. Use the Distributive property to write this as $2AT = AM$. Now multiply each side of the equation by $\frac{1}{2}$ to get $AT = \frac{1}{2}AM$.

Reteaching with Additional Practice

1.

Statements	Reasons
1. $AD = AB = 12$	**1.** Given
2. $\overline{AD} \cong \overline{AB}$	**2.** Def. of congruent segments
3. $\overline{AD} \cong \overline{CD}$, $\overline{CD} \cong \overline{BC}$	**3.** Given
4. $\overline{AB} \cong \overline{BC}$	**4.** Transitive prop. of congruence

2. 3 **3.** 3

Cooperative Learning Activity

1. Given: C is the bisector of $\overline{AB}$.

Prove: $AC = CB$

Statements	Reasons
1. C is the bisector of $\overline{AB}$.	**1.** Given
2. C is the midpoint of $\overline{AB}$.	**2.** Def. of bisector
3. $AC = CB$	**3.** Def. of midpoint

2. You are given that C is the midpoint of $\overline{AB}$. By the definition of midpoint, $AC = CB$. By the segment addition postulate, $AC + CB = AB$. By the substitution property of equality, $AC + AC = AB$. By the distribution property, $2AC = AB$. **3.** *Sample answer:* Continually bisecting one inch is convenient and measurable. Doing this creates increments of 2, 4, 8, 16.

Interdisciplinary Application

1. 16 rods **2.** Transitive Property of Segment Congruence **3.** 6 cuts **4.** 150 inches

5. 15 rods **6.** 14 inches **7.** 3

8. Yes; two pieces; one is 1 inch in length and the other is 2 inches in length

Challenge: Skills and Applications

1. $DE = 14, AE = 35$ **2.** $AB = 3, AE = 15$

3. $C = 8, D = 12, E = 20$

4. $A = 6, C = 0, D = -3$

5. $UX = 4, XY = 6$ **6.** $XY = 15, UV = 40$

7. $U = 3, M = 11, V = 19$

8. $X = -4, Y = -13, V = -22$

9. $NO = 8, MO = 17$

10. $MO = 53, MQ = 106$

11. $PQ = 56, MQ = 130$

12. *Sample answer:*

Statements	Reasons
1. $\overline{BD} \cong \overline{AC}$	**1.** Given
2. $BD = AC$	**2.** Def. of $\cong$ segments
3. $BD = 12$	**3.** Given
4. $AC = 12$	**4.** Transitive prop. of equality
5. $CD = x, AD = 13$	**5.** Given
6. $\angle C$ is a right angle	**6.** Given
7. $\triangle ACD$ is a right $\triangle$	**7.** Def. of right $\triangle$
8. $(AC)^2 + (CD)^2 = (AD)^2$	**8.** Pythagorean thm.
9. $12^2 + x^2 = 13^2$	**9.** Substitution prop. of equality
10. $144 + x^2 = 169$	**10.** Simplify.
11. $x^2 = 25$	**11.** Subtraction prop. of equality

Lesson 2.6

Lesson 2.6

Warm-up Exercises

1. 90° **2.** 48° **3.** 138°
4. *Sample answer:* 45°, 45°
5. *Sample answer:* 90°, 90°

Daily Homework Quiz

1.

Statements	Reasons
1. $\overline{AB} \cong \overline{BC}$, $\overline{CD} \cong \overline{BC}$	**1.** Given
2. $\overline{AB} \cong \overline{CD}$	**2.** Transitive Prop. of Segment Cong.
3. $AB = CD$	**3.** Def. of ≅ segments
4. $2m + 6 = 3m - 1$	**4.** Substitution prop. of equality
5. $-m + 6 = -1$	**5.** Subtraction prop. of equality
6. $-m = -7$	**6.** Subtraction prop. of equality
7. $m = 7$	**7.** Division prop. of equality

2. Check students' work.

Lesson Opener

Allow 10 minutes.

1. Check work. **2.** ∠2 or ∠5
3. ∠2 and ∠5 are congruent. **4.** ∠3 or ∠4
5. ∠3 and ∠4 are congruent.

Practice A

1. false **2.** true **3.** false 4. true

5.

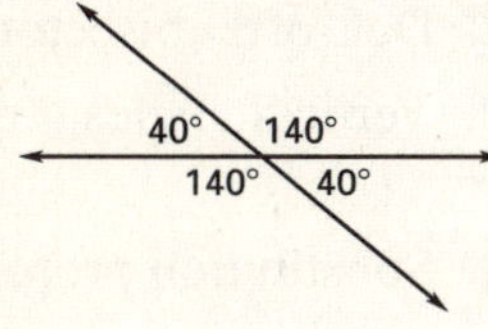

6.

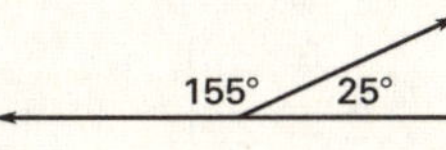

7.

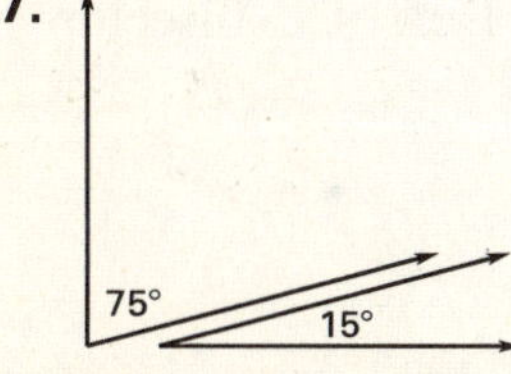

8.

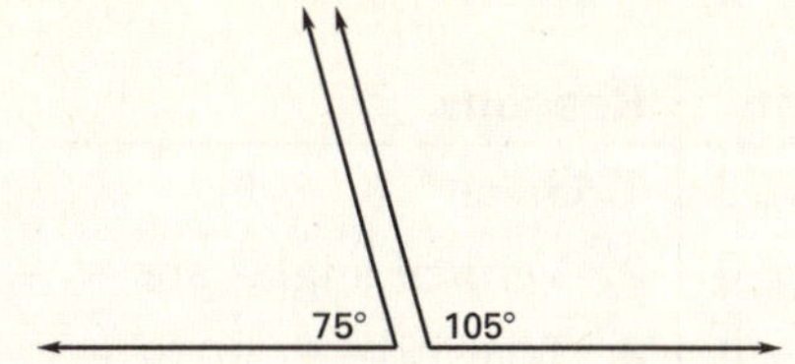

9. $x = 40$ **10.** $x = 50$ **11.** $x = 15$

12. 2. Definition of complementary angles
4. Definition of congruent angles
5. Substitution property of equality
6. Substitution property of equality
7. Definition of complementary angles

13.

Statements	Reasons
1. ∠6 ≅ ∠7	**1.** Given
2. ∠7 ≅ ∠8	**2.** Vertical angles are ≅
3. ∠6 ≅ ∠8	**3.** Transitive Prop. of ≅
4. ∠5 ≅ ∠6	**4.** Vertical angles are ≅
5. ∠5 ≅ ∠8	**5.** Transitive Prop. of ≅

Practice B

1. false **2.** true **3.** false **4.** true

5.

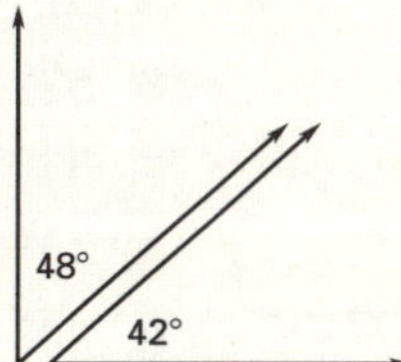

6.

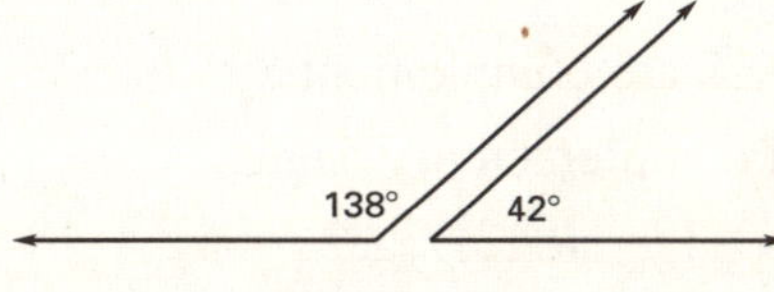

7.

8.

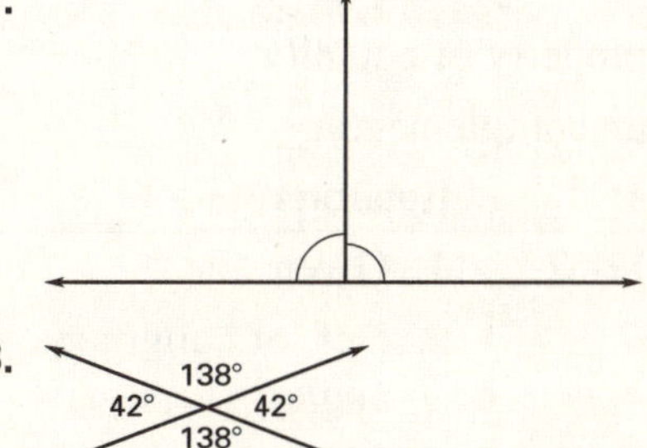

9. $x = 13$ **10.** $x = 17$ **11.** $x = 28$

12. 2. Definition of complementary angles
4. Definition of congruent angles
5. Substitution property of equality
6. Substitution property of equality
7. Definition of complementary angles

Lesson 2.6 *continued*

13.

Statements	Reasons
1. $\angle 2 \cong \angle 3$	1. Given
2. $\angle 3 \cong \angle 4$	2. Vertical angles are $\cong$
3. $\angle 2 \cong \angle 4$	3. Transitive Prop. of $\cong$
4. $\angle 1 \cong \angle 2$	4. Vertical angles are $\cong$
5. $\angle 1 \cong \angle 4$	5. Transitive Prop. of $\cong$

Practice C

1.

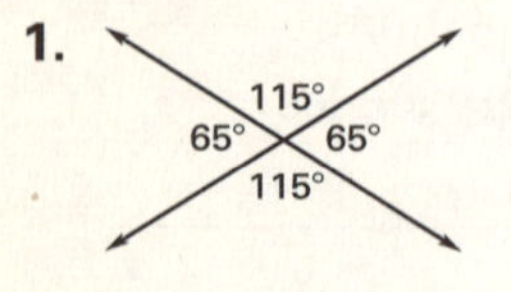

2.

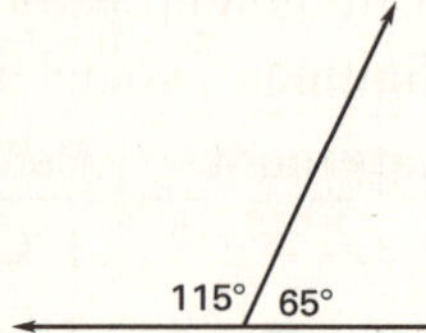

3.

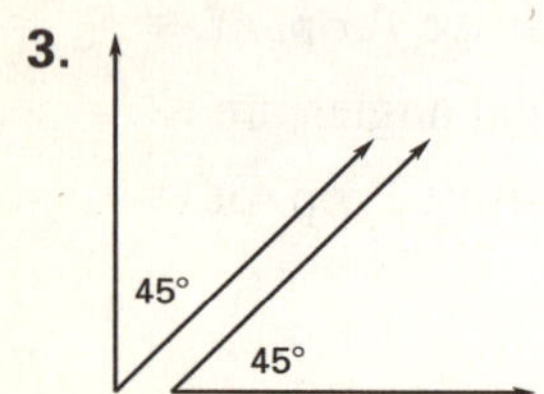

4.

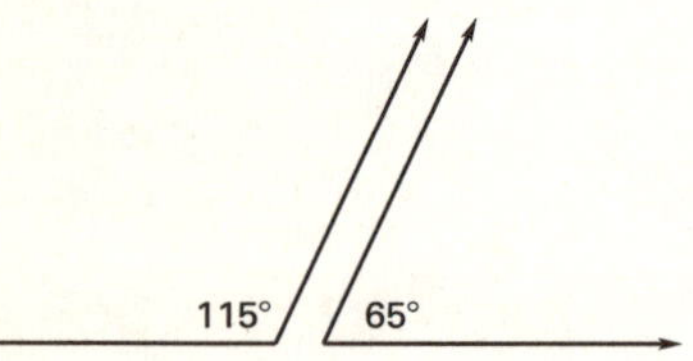

5. 42°
6. 52°
7. 37°
8. 42°
9. 39°

10. 27°

11. 1. $\angle 1$ and $\angle 2$ are complementary
2. Definition of complementary angles
3. Given 4. $m\angle 1 = m\angle 3, m\angle 2 = m\angle 4$
5. Substitution property of equality
6. Substitution property of equality
7. $\angle 3$ and $\angle 4$ are complementary

12.

Statements	Reasons
1. $m\angle 6 = m\angle 7$	1. Given
2. $\angle 6 \cong \angle 7$	2. Def. of congruent angles
3. $\angle 7 \cong \angle 8$	3. Vertical angles are congruent.
4. $\angle 6 \cong \angle 8$	4. Transitive Property of Congruence
5. $\angle 5 \cong \angle 6$	5. Vertical angles are congruent.
6. $\angle 5 \cong \angle 8$	6. Transitive Property of Congruence

13. You are given $m\angle 6 = m\angle 7$, so they are congruent by definition. Since $\angle 7$ and $\angle 8$ are vertical angles, they are congruent. Use the Transitive Property to show $\angle 6 \cong \angle 8$. Now $\angle 5 \cong \angle 6$ since they are vertical angles and finally use Transitive Property to show $\angle 5 \cong \angle 8$.

Reteaching with Additional Practice

1. 90° **2.** 90° **3.** 31° **4.** 125° **5.** 180°
6. 52° **7.** 90°; 36° **8.** 132°; 48°

Real-Life Application

1. 90° **2.** 90° **3.** no, no **4.** 40°
5. 52°, 38°, 38°

Challenge: Skills and Applications

1. a. *Sample answer:*

Statements	Reasons
1. $m\angle 1 + m\angle 2 = m\angle AOC$	1. $\angle$ Addition Prop.
2. $m\angle AOC = 90°$	2. Def. of $\perp$ lines
3. $m\angle 1 + m\angle 2 = 90°$	3. Transitive prop. of equality
4. $\angle 1$ and $\angle 2$ are complementary	4. Def. of complementary angles

b. *Sample answer:* Consider two adjacent angles, each with measure 135°. The nonshared sides of these angles are perpendicular, but the angles are not complementary. **c.** *Sample answer:* By drawing right angle $\angle AOC$ with B in the interior of the angle, we implicitly assumed that $\angle 1$ and $\angle 2$ are acute.

2. *Sample answer:*

Statements	Reasons
1. $\overrightarrow{TV}$ bisects $\angle UTW$	1. Given
2. $\angle 1 \cong \angle 2$	2. Def. of $\angle$ bisector
3. $\angle 1 \cong \angle 3$, $\angle 2 \cong \angle 4$	3. Vertical angles are $\cong$.
4. $\angle 3 \cong \angle 4$	4. Substitution prop. of equality
5. $\overrightarrow{TY}$ bisects $\angle XTZ$	5. Def. of $\angle$ bisector

Lesson 2.6 *continued*

3. *Sample answer:* Since $\angle 1 \cong \angle 2$, $m\angle 1 = m\angle 2$. But $\angle 1$ and $\angle 2$ form a linear pair, so they are supplementary. Therefore, $m\angle 1 + m\angle 2 = 180°$. But $m\angle 1 = m\angle 2$ (given), so, by substitution, $m\angle 1 + m\angle 1 = 180°$, or $2m\angle 1 = 180°$. By the division property of equality, $m\angle 1 = 90°$, so $\angle 1$ is a right angle. Therefore, by definition of perpendicular lines, $\overleftrightarrow{KM} \perp \overleftrightarrow{LN}$.

4. *Sample answer:* Since $\overrightarrow{MK}$ bisects $\angle PMQ$, $m\angle 2 = m\angle 3$. Since $\angle 1 \cong \angle 4$, $m\angle 1 = m\angle 4$. Therefore, $m\angle 2 + m\angle 1 = m\angle 3 + m\angle 4$. But $m\angle 2 + m\angle 1 = m\angle KML$ and $m\angle 3 + m\angle 4 = m\angle KMN$, so $\angle KML \cong \angle KMN$. Therefore, by the previous exercise, $\overrightarrow{KM} \perp \overrightarrow{LN}$, so $m\angle KML = 90°$. By substitution, $m\angle 2 + m\angle 1 = 90°$, so $\angle 1$ and $\angle 2$ are complementary.

5. a. They are perpendicular.
b. *Sample answer:* Since $\overrightarrow{OB}$ bisects $\angle AOC$ and $\overrightarrow{OD}$ bisects $\angle COE$, $m\angle 1 = m\angle 2$ and $m\angle 3 = m\angle 4$. By the Angle Addition Postulate, $m\angle AOC = m\angle 1 + m\angle 2$ and $m\angle COE = m\angle 3 + m\angle 4$. Therefore, by substitution, $m\angle AOC = 2m\angle 2$ and $m\angle COE = 2m\angle 3$. Since $\angle AOC$ and $\angle COE$ form a linear pair, they are supplementary and $m\angle AOC + m\angle COE = 180°$. By substitution, $2m\angle 2 + 2m\angle 3 = 180°$; applying the division property of equality gives $m\angle 2 + m\angle 3 = 90°$. But $m\angle 2 + m\angle 3 = m\angle BOD$ by the Angle Addition Postulate, so $m\angle BOD = 90°$. Since $\angle BOD$ is a right angle, $\overrightarrow{OB} \perp \overrightarrow{OD}$.

Quiz 2

1. $x + 5 = 7$ (Given)
$x = 2$ (Subtraction property of equality)

2. $x - 6 = -3$ (Given)
$x = 3$ (Addition property of equality)

3. $3x + 4 = 13$ (Given)
$3x = 9$ (Subtraction prop. of equality)
$x = 3$ (Division property of equality)

4. $x + 5 = 2x + 5$ (Given)
$5 = x + 5$ (Subtr. prop. of equality)
$0 = x$ (Subtr. prop. of equality)

5. $2(4x + 2) = 20$ (Given)
$8x + 4 = 20$ (Distributive property)
$8x = 16$ (Subtr. prop. of equality)
$x = 2$ (Division prop. of equality)

6. $-(-3x + 2) = 13$ (Given)
$3x - 2 = 13$ (Distributive property)
$3x = 15$ (Add. prop. of equality)
$x = 5$ (Div. prop. of equality)

7.

Statements	Reasons
1. $\angle AXB \cong \angle CXD$	1. Given
2. $m\angle AXB = m\angle CXD$	2. Def. of congruent angles
3. $m\angle BXC = m\angle BXC$	3. Reflective property
4. $m\angle AXB + m\angle BXC = m\angle CXD + m\angle BXC$	4. Addition property of equality
5. $m\angle AXB + m\angle BXC = m\angle AXC$	5. Angle addition postulate
6. $m\angle CXD + m\angle BXC = m\angle BXD$	6. Angle addition postulate
7. $m\angle AXC = m\angle BXD$	7. Sub. prop. of equality
8. $\angle AXC \cong \angle CXD$	8. Definition of congruent angles

8.

Statements	Reasons
1. $\overline{XY} \cong \overline{ZW}$	1. Given
2. $XY = ZW$	2. Def. of congruent segments
3. $YZ = YZ$	3. Reflexive
4. $XY + YZ = ZW + YZ$	4. Add. prop. of equality
5. $XY + YZ = XZ$	5. Segment addition postulate
6. $ZW + YZ = YW$	6. Segment addition postulate
7. $XZ = YW$	7. Substitution prop. of equality
8. $\overline{XZ} \cong \overline{YW}$	8. Def. of congruent segments

Lesson 2.6 *continued*

9. 55°

Review and Assessment

Test A

1. Inverse: If the sky is not clear, then we cannot see the stars at night. Converse: If we can see the stars at night, then the sky is clear. Contrapositive: If we cannot see the stars at night, then the sky is not clear.

2. Inverse: If I do not go to the beach, then I won't go swimming. Converse: If I go swimming, then I will go to the beach. Contrapositive: If I do not go swimming, then I do not go to the beach.

3. Inverse: If the point is not on the line, then it is not in the plane. Converse: If the point is in the plane, then it is on the line. Contrapositive: If the point is not in the plane, then it is not on the line.

4. Inverse: If we do not use a calculator, then we are not in math class. Converse: If we are in math class, then we use a calculator. Contrapositive: If we are not in math class, then we do not use a calculator.

5. false **6.** true

7. If the angle has a measure of 90°, then the angle is a right angle.

8. The angle is not a right angle.

9. The angle does not have a measure of 90°.

10. If the angle is a right angle, then the angle has a measure of 90°.

11. If the angle does not have a measure of 90°, then the angle is not a right angle.

12. If the angle is not a right angle, then the angle does not have a measure of 90°.

13. Subtraction property of equality

14. Division property of equality

15. Substitution property of equality

16. Reflexive property of equality

17. Addition property of equality

18. 11 **19.** 3

20. 1. Linear Pair Postulate
2. Substitution property of equality
3. Subtraction property of equality

21. 1. Given 2. Segment Addition Postulate
3. Substitution property of equality
4. Distributive property
5. Division property of equality

Test B

1. Inverse: If the sky is not clear, then we cannot see the stars at night. Converse: If we can see the stars at night, then the sky is clear. Contrapositive: If we cannot see the stars at night, then the sky is not clear.

2. Inverse: If I do not ride on the Ferris wheel, then I am afraid of heights. Converse: If I am not afraid of heights, then I will ride on the Ferris wheel. Contrapositive: If I am afraid of heights, then I do not ride on the Ferris wheel.

3. Inverse: If the point is not on the line, then it is not in the plane. Converse: If the point is in the plane, then it is on the line. Contrapositive: If the point is not in the plane, then it is not on the line.

4. Inverse: If we do not use a calculator, then we are not in math class. Converse: If we are in math class, then we use a calculator. Contrapositive: If we are not in math class, then we do not use a calculator.

5. true **6.** true

7. If points *X*, *Y*, and *Z* are noncollinear, then points *X*, *Y*, and *Z* are coplanar.

8. Points *X*, *Y*, and *Z* are noncoplanar.

9. Points *X*, *Y*, and *Z* are collinear.

10. If points *X*, *Y*, and *Z* are coplanar, then points *X*, *Y*, and *Z* are noncollinear.

11. If points *X*, *Y*, and *Z* are collinear, then points *X*, *Y*, and *Z* are noncoplanar.

12. If points *X*, *Y*, and *Z* are noncoplanar, then points *X*, *Y*, and *Z* are collinear.

13. Subtraction property of equality

14. Division property of equality

15. Substitution property of equality

16. Reflexive property of equality

17. Addition property of equality

18. 3 **19.** 14

Review and Assessment *continued*

20.

Statements	Reasons
1. $AB = BC$	**1.** Given
2. $AC = AB + BC$	**2.** Segment Addition Postulate
3. $AC = BC + BC$	**3.** Substitution property of equality
4. $AC = 2BC$	**4.** Distributive property
5. $\frac{1}{2}AC = BC$	**5.** Division property of equality

21.

Statements	Reasons
1. $m\angle 1 + m\angle 3 = 180°$ $m\angle 2 + m\angle 3 = 180$	**1.** Linear Pair Postulate
2. $m\angle 1 + m\angle 3 = m\angle 2 + m\angle 3$	**2.** Substitution prop. of equalty
3. $m\angle 1 = m\angle 2$	**3.** Subtraction prop. of equality

Test C

1. Inverse: If the sky is not clear, then we cannot see the stars at night. Converse: If we can see the stars at nght, then the sky is clear. Contrapositive: If we cannot see the stars at night, then the sky is not clear. **2.** Inverse: If I do not sail on a boat, then I will not get seasick. Converse: If I get seasick, then I am sailing on a boat. Contrapositive: If I do not get seasick, then I am not sailing on a boat. **3.** Inverse: If the angle is not less than 90°, then it is not acute. Converse: If the angle is acute, then it is less than 90°. Contrapositive: If the angle is not acute, then it is not less than 90°.

4. Inverse: If the tree is not an evergreen, then it loses its leaves in the fall. Converse: If the tree does not lose its leaves in the fall, then the tree is an evergreen. Contrapositive: If the tree loses its leaves in the fall, then the tree is not an evergreen.

5. If two segments are congruent, then they have the same length. **6.** If an angle is a right angle, then it has a measure of 90°. **7.** If points *X*, *Y*, and *Z* are noncollinear, then points *X*, *Y*, and *Z* form a triangle. **8.** Points *X*, *Y*, and *Z* do not form a triangle. **9.** Points *X*, *Y*, and *Z* are collinear.

10. If points *X*, *Y*, and *Z* form a triangle, then the points *X*, *Y*, and *Z* are noncollinear. **11.** If points *X*, *Y*, and *Z* are collinear, then points *X*, *Y*, and *Z* do not form a triangle.

12. If points *X*, *Y*, and *Z* do not form a triangle, then points *X*, *Y*, and *Z* are collinear.

13. Subtraction property of equality

14. Division property of equality

15. Substitution property of equality

16. Reflexive property of equality (symm)

17. Addition property of equality

18. 4 **19.** 6

20.

Statements	Reasons
1. $\overline{AB} \cong \overline{CD}, \overline{CD} \cong \overline{EF}$	**1.** Given
2. $AB = CD, CD = EF$	**2.** Definition of congruent segments
3. $AB = EF$	**3.** Transitive property of equality
4. $\overline{AB} \cong \overline{EF}$	**4.** Definition of congruent segments

21.

Statements	Reasons
1. $\angle 2 \cong \angle 3, \angle 4 \cong \angle 5$	**1.** Given
2. $\angle 3 \cong \angle 4, \angle 1 \cong \angle 2$	**2.** Vertical angles are congruent.
3. $\angle 1 \cong \angle 3, \angle 3 \cong \angle 5,$ $\angle 1 \cong \angle 5$	**3.** Transitive property of congruence
4. $m\angle 5 + \angle 6 = 180°$	**4.** Linear Pair Postulate
5. $m\angle 1 + \angle 6 = 180°$	**5.** Substitution prop. of equality
6. $\angle 1$ is supp to $\angle 6$	**6.** Definition of supplementary angles

SAT/ACT Test

1. D **2.** B **3.** C **4.** B **5.** A **6.** B **7.** A

Alternative Assessment

1. Complete answers should include:

- If it is Saturday, then you have to work at 9 A.M.
- If you have to work at 9 A.M., then it is Saturday. • $\sim q \rightarrow \sim p$ • $\sim p \rightarrow \sim q$

Review and Assessment *continued*

• If it is Saturday, then you have to wake up at 7:30 A.M. **2.** Given; Segment Addition Postulate; Segment Addition Postulate; Addition property of equality; Substitution property of equality

3. a.

Statements	Reasons
1. J is the midpoint of $\overline{AB}$.	**1.** Given
2. $AJ = JB$	**2.** Def. of a midpoint
3. $AJ + JB = AB$	**3.** Segment Addition Postulate
4. $AJ + AJ = AB$	**4.** Substitution prop. of equality
5. $2 \cdot AJ = AB$	**5.** Distributive prop.
6. $AJ = \frac{1}{2}AB$	**6.** Division property of equality

b.

Statements	Reasons
1. $\angle 1 \cong \angle 2$ and $\angle 2 \cong \angle 3$	**1.** Given
2. $m\angle 1 \cong m\angle 2$ and $m\angle 2 \cong m\angle 3$	**2.** Def. of $\cong$ ∠s
3. $m\angle 1 \cong m\angle 3$	**3.** Transitive Prop. of $\cong$
4. $\angle 1 \cong \angle 3$	**4.** Def. of $\cong$ ∠s

4. Answers may vary.

Project: Investigating Advertising Slogans

1. Check that student includes a source. *Sample answer:* Cereal *XYZ* makes you grow big and strong. **2.** *Sample answer:* If you eat Cereal *XYZ*, then you will grow big and strong.

3. *Sample answer:* Inverse: If you do not eat Cereal *XYZ*, you will not grow big and strong. Converse: If you are growing big and strong, then you are eating Cereal *XYZ*. Contrapositive: If you are not growing big and strong, then you are not eating Cereal *XYZ*.

4. Check answers. Look for analysis of the relationship between the different conditional statements. The contrapositive is logically equivalent to the original conditional but no statement is true just because it can be written as a conditional statement. **5.** Check answers.

Cumulative Review

1. 7.81 **2.** 12.53 **3.** 7.28 **4.** 5 **5.** (5, 2) **6.** $\left(\frac{5}{2}, \frac{7}{2}\right)$ **7.** $\left(\frac{1}{2}, -3\right)$ **8.** $\left(\frac{3}{2}, 2\right)$ **9.** 28° **10.** 28° **11.** 62° **12.** 118° **13.** 80°, 100° **14.** 47°, 133° **15.** 60°, 120° **16.** 55°, 125° **17.** 20 ft **18.** 31.4 ft **19.** inverse: If two angles are not right angles, then they are not congruent; converse: If two angles are congruent, then they are right angles; contrapositive: If two angles are not congruent, then they are not right angles.

20. inverse: If $x \neq 2$, then $x^2 \neq 4$; converse: If $x^2 = 4$, then $x = 2$; contrapositive: If $x^2 \neq 4$, then $x \neq 2$.

21. conditional statement: If two angles are complementary, then their measures add to 90°; converse: If two angle measures add to 90°, then they are complementary.

22. conditional statement: If three points are collinear, then they lie on the same line; converse: If three points lie on the same line, then they are collinear. **23.** p: It is snowing; q: Jennifer will ride the bus to school; $\sim p \rightarrow \sim q$: If it is not snowing then Jennifer will not ride the bus to school; $\sim q \rightarrow \sim p$: If Jennifer does not ride the bus to school, then it is not snowing.

24. p: The car runs out of gas; q: It will not run; $\sim p \rightarrow \sim q$: If the car does not run out of gas, then it will run; $\sim q \rightarrow \sim p$: If the car runs, then it is not out of gas.

25. Division property of equality

26. Subtraction property of equality

27. Symmetric property of equality

28. Substitution property of equality

29. Multiplication property of equality

30. Addition property of equality

31. (a) Given (b) Linear Pair Postulate (c) Linear Pair Postulate (d) Congruent Supplements Theorem

32. (a) Given (b) Addition property of equality

(c) Addition property of equality

(d) Reflexive property of equality

(e) Substitution property of equality

(f) Subtraction property of equality

(g) Definition of congruent segments